TOOLS AND WEAPONS

TOOLS AND WEAPONS

The Promise and the Peril
of the Digital Age

BRAD SMITH

and CAROL ANN BROWNE

PENGUIN PRESS

NEW YORK

2019

PENGUIN PRESS
An imprint of Penguin Random House LLC
penguinrandomhouse.com

Copyright © 2019 by Bradford L. Smith

ISBN 9781984877710 (hardcover)
ISBN 9781984877727 (ebook)
ISBN 9781984879226 (export)

Printed in the United States of America
1   3   5   7   9   10   8   6   4   2

DESIGNED BY AMANDA DEWEY

TO OUR MOTHERS:

Neither is here to read this,
but their appreciation for books
helped inspire us to write it

CONTENTS

Foreword

I first turned to Brad Smith for advice during the toughest time in my professional life. Two decades later, I haven't stopped.

Brad joined Microsoft's legal team in 1993, but we really got to know each other in the late 1990s, during the US government's antitrust suit against the company. We spent countless hours working side by side. I could see right away what a sophisticated thinker he was. I came to like Brad as a person and to trust his judgment as a professional.

Brad shaped our legal strategy during the lawsuit, and then he did something else that was at least as important: He ushered in a big cultural and strategic shift at the company. That shift is at the heart of this book.

In the early days of Microsoft, I prided myself on how little time we spent talking to people in the federal government. I would tell people, "Isn't it great that we can be successful and not even have an office in DC?" As I learned the hard way during the antitrust suit, this was not a wise position to take.

After the case was settled, Brad persuaded me and a lot of other people at Microsoft that we needed to take a different approach. Then he showed us how to make it happen. Brad is a lawyer, not a software developer, and although he has a great command of technology, he didn't think quite the same way as the rest of us. (I mean this as a compliment.) He saw that we needed to put more time and energy into connecting with different constituencies, including the government, our partners, and sometimes even our competitors. Brad would have made a great diplomat, which makes sense given his early interest in international relations.

It says a lot about Brad that his thinking was not limited to Microsoft's own self-interest. He recognized the central importance of technology and the policies that affect it, and he concluded that staying on the sidelines wasn't just a mistake for our company—it was a mistake for the industry. Although there would be times when we would need to go it alone, there would be many others—for example, when artificial intelligence, facial recognition, and cybersecurity are involved—when we all have much more to gain from working with one another.

As he argues in this book, there are also times when it is in everyone's interest for the government to step in with more regulation. (Brad is self-aware enough to see the irony of a business leader asking for more government rules, rather than fewer.) To that end, he knew that Microsoft and other tech companies needed to engage more with leaders in the United States, Europe, and elsewhere. My days of bragging about not having an office in DC were over.

Brad's vision has never been more relevant. Governments around the world are taking a hard look at many technology companies, and at the industry as a whole. How is their technology being used? What impact is it having? What responsibilities do tech companies have? How should governments and the broader community think about these issues?

Although these are not the same questions we faced twenty years ago, the insights that Brad had back then are just as valuable today.

Take, for example, the issues raised by facial-recognition technology. This isn't yet a big topic for public debate, but it will be. What limits should software companies put on the use of their facial recognition tools? How should the industry think about this, and what kind of government regulations make sense?

Brad has led the way in anticipating these questions and creating partnerships to discuss them. The tech industry will need to come together, working with customers and governments around the world. It may not be possible to get everyone on board, but if we let things fragment so that the rules vary hugely from country to country, it won't be good for customers, the tech industry, or society.

Tools and Weapons covers an impressive range of fifteen issues, including cybersecurity, the diversity of the IT workforce, and the relationship between the United States and China. If I had to pick the most important chapter, it would be the one on privacy. The ability to collect huge amounts of data is a double-edged sword. On the one hand, it empowers governments, businesses, and individuals to make better decisions. On the other hand, it raises big questions about how we can use that data while protecting people's right to privacy.

Yet, as Brad shows, while the technology is relatively new, the questions themselves are not. People have been wrestling with one version or another of this problem for centuries. Although you may expect a chapter about data privacy to touch on the way Nazi Germany collected information on its own people, you might not expect it to also mention the War of 1812 or give you a quick history of Mutual Legal Assistance Treaties.

That shows you Brad's wide-ranging interests and his ability to go deep on almost any subject. But none of it comes off as a dull legal brief. Brad and his coauthor, Carol Ann Browne, are great storytellers who give you an insider's view of what it is like to hammer out these issues in real time, in conference halls and courtrooms around the world. Brad is not just sitting back and analyzing things—he is bringing people together to find solutions.

Brad and I are in constant contact about these issues, in person and over email. I still rely on his wisdom and judgment today. Given his experience and intelligence, you could not ask for a better guide to thinking through the questions facing the technology industry now.

These issues are becoming only more important. *Tools and Weapons* offers a clear view of the questions raised by new technologies, and a potential path forward for tech companies and for society. Brad has written a clear, compelling guide to some of the most pressing debates in technology today.

Bill Gates
April 2019

⟩⟩ **THE CLOUD:**
The World's Filing Cabinet

C ivilization has always run on data.

Human history began when people developed the ability to speak. With the invention of language, people could share their ideas, experiences, wants, and needs.

Progress accelerated as people developed the ability to write. Ideas spread more easily and accurately not just from person to person, but from place to place.

Then came the spark that created the flame of knowledge: the ability to store, retrieve, and share what people had written. A hallmark of the ancient world became the construction of libraries.[1] These archives of documents and books meant that people could communicate more readily not just across space, but across time, passing information from one generation to the next. Centuries later, when Johannes Gutenberg invented the mechanical printing press, the flame became a fire that empowered writers and readers alike.

That fire would sweep the world. The ensuing centuries produced an explosion in commerce that was both cause and consequence of increasing communication. By the early twentieth century, every office needed a facility to store documents. Rooms were filled with filing cabinets.[2]

While data has always been important to society, it has never played the role it does today. Even when trade slows or economies falter, data continues to grow at a steady pace. Some say that data has become the oil of the twenty-first century. But that understates the reality. A century ago, automobiles, airplanes, and many trains ran on oil. Today, every aspect of human life is fueled by data. When it comes to modern civilization, data is more like the air we breathe than the oil we burn.

Unlike oil, data has become a renewable resource that we humans can create ourselves. This decade will end with almost 25 times as much digital data as when it began.[3] With artificial intelligence, or AI, we're doing more with data than ever before.

We call the digital infrastructure that supports this the cloud. While its name sounds soft and fluffy, in truth the cloud is a fortress. Every time you look something up on your smartphone, you pull data from a mammoth data center—a modern-day marvel that almost no one gets to step inside.

But if you're lucky enough to visit a data center, you'll better understand how the world now works.

One of the best places to see the inner workings of the cloud is the world's capital of apples. The tiny town of Quincy, Washington, sits roughly 150 miles east of Seattle off Interstate 90. Its location is no accident. Quincy is in the center of the state's agricultural basin, perched near a steep gorge carved for millennia by the wide and rolling Columbia River, the largest waterway in the western United States. The town is powered by a network of hydroelectric plants, including the Grand Coulee Dam, the largest power station in the United States. It's an ideal setting for what has become the world's biggest consumer of electricity, the modern data center.[4]

A few blocks off Quincy's main street, you'll find a series of non-descript buildings secured by tall fences and walls. Some are marked by the logos of today's tech companies; others have no identification at all. The largest of these facilities is called the Columbia Data Center, which is owned by Microsoft.

It's thrilling—and a bit eerie—to take in the sheer size of a data center. Our facilities in Quincy are no longer just a single building. They fill two data center campuses with more than twenty buildings, totaling two million square feet. Each building is the size of two football fields and is big enough to house two large commercial airplanes. This collection of buildings is home to hundreds of thousands of server computers and millions of hard disks, each of which is replaced with faster, more efficient models every three years.

The best way to get a sense of the scale of a data center is to walk from its outer edge to its center. Sitting outside the walls of each building are some of the world's largest electrical generators, ready to power up within seconds to ensure the data center doesn't skip a beat if the region's electrical grid goes down. Each generator stands more than twenty feet high and can power the equivalent of more than two thousand homes. The generators are connected to diesel fuel tanks that can keep the data center running off the grid for forty-eight hours, with refueling arrangements in place to keep the operations up well beyond if ever needed. In our newer operations, like the one in Cheyenne, Wyoming, the generators run on cleaner natural gas and provide backup power to the region's grid. Dozens of these huge generators stand next to the data center buildings, ready in case there is a local outage from the hydroelectric power supplied by the Grand Coulee Dam.

Inside each building, a string of large secure rooms operate as electrical substations, typically pulling power from the electrical grid at 230,000 volts before reducing it to 240 volts to power the data center's computers. These substation rooms are lined with rows of six-foot racks, each connecting five hundred or more batteries that look like what you'd find under the hood of your car. Every door to the room is bulletproof and every wall is fireproof, so a fire can't spread from one room to another. A typical data center building has four or more of these rooms, and depending on the configuration, the building may house as many as five thousand batteries. They serve two purposes. First, electricity from the grid flows through the racks, keeping the batteries charged and circumventing a potential electrical spike, so the flow of electricity to the computers remains smooth and constant. And in the event of a power outage, the batteries will keep the data center operating until the generators start up.

Through another set of bulletproof doors and fireproof walls, another airport-style metal detector manned by two uniformed guards stands between you and the inner sanctum of the complex. Only full-time Microsoft employees with their names on a preapproved list can go any farther. As you enter a small receiving room, a steel door closes behind you. You wait, locked inside as security staff look you over through a camera before opening the next bulletproof door.

Finally, you enter a cavernous room, a temple to the information age and cornerstone of our digital lives. A hushed hum welcomes you into the nerve center containing floor-to-ceiling racks filled with computers, lined up in a formation that extends beyond your line of sight. This massive library of steel and circuits contains servers each identical in footprint but containing its own unique volume of data. It is the digital world's filing cabinet.

Somewhere in one of these rooms in one of these buildings, there are data files that belong to you. They have the email you wrote this morning, the document you worked on last night, and the photo you took yesterday afternoon. They also likely contain personal information from your bank, doctor, and employer. The files occupy just a tiny sliver of a hard drive on one of these thousands and thousands of computers. Each file is encrypted, meaning that the information is encoded so that only authorized users of that data can read it.

Each data center building contains multiple rooms like this, sealed off from one another in case of fire. Each set of computers is connected to three power sources within the building. And each row is designed to recirculate throughout the building the heat given off by its computers to reduce the need for heating and hence electrical demand in the winter.

As you leave the server room, you endure the entire security routine yet again. Off come your shoes and then your belt. Just as you pause and consider the fact that you don't have to endure this to *leave* an airport, your host reminds you that there is security in both directions for a reason. Microsoft wants to ensure that no one can copy data on a thumb drive or steal a hard disk with someone's personal data. Even the hard disks themselves leave through a special exit. When it's time to replace them, their data is copied to a new computer and the files are erased. Then the retired hard disk goes through the huge metal equivalent of a shredding machine.

In some ways, the most remarkable feature is saved for the end of the tour. Your guide explains that each data center region has another set of buildings like this one, so the data of a business, government, or nonprofit can constantly be backed up somewhere else. This way, if there is an earthquake, hurricane, or some other natural or man-made disaster, the second data center will step in to keep your cloud service operating smoothly. As we found when an earthquake rocked northern Japan, our data center in southern Japan ensured there was no interruption of service.

Today Microsoft owns, operates, and leases data centers of all sizes in more than 100 locations in more than 20 countries (and growing), delivering 200 online services and supporting more than a billion customers in more than 140 markets.

When I joined Microsoft in 1993, one didn't need much capital to start a software company. Bill Gates and Paul Allen, our two cofounders, were the latest in a line of tech developers who had launched their companies from a garage or a college dorm room. The point was that the creation of software didn't require a lot of money. A good computer, a small savings account, and a willingness to eat a lot of pizza were enough to get you started.

We saw this again and again as Microsoft grew from a tiny start-up to the multinational it is today. In 2004, we looked to acquire an anti-spyware software company called Giant Company Software. When our team reached out to them, we called the published telephone number that provided technical support. When the caller from Microsoft asked to be transferred to the company's CEO, the phone was simply handed to the only other employee, who happened to sit across the desk. Not surprisingly, the acquisition negotiation progressed quickly.[5]

I can't help but think about Giant Company Software when I visit one of our data centers. You can still create a new software app the way Bill and Paul got started. Open-source developers do it all the time. But providing the platforms needed for cloud computing at a global scale? That's a different story. As I walk among the thousands of blinking computers, racks of batteries, and enormous generators, it feels like more than a different era. It seems like a different planet. Data center campuses cost hundreds of millions of dollars to build. And once construction ends, the work to maintain and upgrade the facility begins. Sites are expanded, and servers, hard drives, and batteries are upgraded or swapped for newer and more efficient equipment. A data center is never done.

In many ways, the modern data center sits at the center of the new digital era that the world has entered. Its massive accumulation of data, storage, and computing power has created an unprecedented platform for progress across the economies of the world. And it has unleashed many of the most challenging issues of our time. How do we strike the right balance between public safety, individual convenience, and personal privacy in this new era? How do we protect ourselves from cyberattacks that are using this technology to disrupt our countries, businesses, or personal lives? How do we manage the economic effects that are now rippling across our communities? Are we creating a world that will have jobs for our children? Are we creating a world we can even control?

The answers to all these questions need to start with a better appreciation for how technology is changing, based in part on understanding how it has changed in the past.

Since the dawn of time, any tool can be used for good or ill. Even a broom can be used to sweep the floor or hit someone over the head. The more powerful the tool, the greater the benefit or damage it can cause. While sweeping digital transformation holds great promise, the world has turned information technology into both a powerful tool and a formidable weapon.

Increasingly this new technology era has ushered in a new age of anxiety. This tension is the most pronounced in the world's democracies. Wracked by unease about immigration, trade, and income inequality, these nations increasingly confront populist and nationalist fissures that result in part from seismic technology shifts. Technology's benefits aren't distributed evenly, and the nature and speed of change is challenging individuals, communities, and entire nations.

Democratic societies collectively confront greater challenges than they've faced in almost a century, and in some cases other countries are using technology itself to exploit this vulnerability.

This book examines these issues from the cockpit of one of the world's largest technology companies. It tells the story of a tech sector trying to come to terms with forces that are bigger than any one company or even the entire industry. In so doing, it tells a story not just about trends and ideas, but about people, decisions, and actions to address a rapidly changing world.

It's an unfolding drama that we at Microsoft sometimes see from a different vantage point. Two decades ago, we were thrust into the heart of what might be considered modern information technology's first collision with the world. In the United States, the Department of Justice and twenty states brought an antitrust lawsuit and sought to break Microsoft into pieces. Governments in other countries followed with their own cases. Competition officials concluded that the Windows operating system was too important to be left unregulated.

While we successfully defended against the breakup of the company, it was a difficult, bruising, and even painful experience. When I was appointed the company's general counsel in 2002, it became my job to hammer out the equivalent of peace treaties with governments around the world and companies across the tech sector. It took almost a decade,[6] and we made more than our share of mistakes. Given my role, I personally was responsible in some way for almost all of them.

But we emerged from the challenges both older and wiser. We learned that we needed to look in the mirror and see what others saw in us and not just what we wanted to see in ourselves. It was like being in the first class to graduate from a new school. We weren't necessarily first in the class, but we had the benefit of getting through school before everyone else.

Today's technology issues are far broader and deeper than they were twenty years ago. We've reached a critical inflection point for both technology and society—a time that beckons with opportunity but that also calls for urgent steps to address pressing problems.

As a result, like Microsoft two decades ago, the tech sector will need to change. The time has come to recognize a basic but vital tenet: When your technology changes the world, you bear a responsibility to help address the world that you have helped create. This might seem uncontroversial, but not in a sector long focused obsessively on rapid growth, and sometimes on disruption as an end in itself. In short, companies that create technology must accept greater responsibility for the future.

But another tenet is equally important: The tech sector cannot address these challenges by itself. The world needs a mixture of self-regulation and government action. Here too there are heightened implications for the world's democracies, in part because they are the most dependent on sustaining a broad economic and social consensus at a time when technology is such a disruptive force. More than ever, it seems difficult for many democratic governments to summon the will to act. But this is a time when democratic governments must move forward with new policies and programs—separately, with each other, and in a new form of collaboration with the tech sector itself. Put simply, governments need to move faster and start to catch up with the pace of technology.

These challenges come without a playbook, but there are nonetheless important insights that can be learned and applied from the past. Technological change has rippled around the world in multiple waves since the first industrial revolution began in England's

Midlands in the mid-1700s. For every current challenge that seems unprecedented, there is often a historical counterpart that, while distinct, has insights for our day. This book speaks to the opportunities and challenges of the future in part by drawing upon the lessons of the past—with thoughts about how we can learn from them.

Ultimately, these questions involve technology and its implications for our jobs, our security, and the world's most fundamental human rights. We need to reconcile an era of rapid technological change with traditional and even timeless values. To achieve this goal, we must ensure that innovation continues but does so in a way that makes technology and the companies that create it subject to democratic societies and our collective capacity to define our destiny.

Chapter 1 ⟫ # SURVEILLANCE: A Three-Hour Fuse

As the early summer sun broke free from the clouds on June 6, 2013, in Redmond, Washington, Dominic Carr twisted the blinds a bit wider in his fifth-floor office on the Microsoft campus. While summer in the Pacific Northwest wouldn't arrive for a full month, the sun spilling through his window was a welcome tease to the warmer days—and slightly slower pace—to come.

He grabbed his phone and headed down the elevator to buy a sandwich from the company café next door. As he walked the busy path between buildings, his phone buzzed in his back pocket. Dominic led the public affairs and communications team reporting to me, handling some of the company's thorniest issues with the media. He was never without his phone—and seldom far from his desk.

An email notification—"Microsoft/PRISM"—lit up his screen. At the time "PRISM" was what we called the company's annual gathering of sales leaders. Just another routine communication about routine Microsoft business.

But this was no everyday email. It was a three-hour fuse on an issue that would soon explode around the globe.

"We're writing to give you notification that the *Guardian* is preparing for publication this evening an article regarding PRISM—a voluntary, secret program of cooperation between several large US technology companies and the NSA," the email began, referring to the National Security Agency in the United States.

The email came from another Dominic—Dominic Rushe—a reporter for the British daily the *Guardian*. It had initially landed in the inbox of a Microsoft public relations manager in Boston, who forwarded it with what we call a "red bang"—an email tag with an exclamation mark essentially saying, "You need to look at this now."

The message included a complex nine-point list for comment and imposed an impossible deadline. Rushe explained, "as responsible journalists, we would like to give you the opportunity to address any specific inaccuracies contained in the above numbered points. . . . We have already approached the White House with regard to this story. Given the sensitive nature of the program this is the earliest opportunity we have had to contact you for comment." He wanted a response by six p.m. Eastern Daylight Time, or three p.m. for Seattle.

The *Guardian* had obtained classified intelligence documents that detailed how nine US technology companies—Microsoft, Yahoo, Google, Facebook, Paltalk, YouTube, Skype, AOL, and Apple—purportedly had signed up for a voluntary program, called PRISM, granting the NSA direct access to email, chat, video, photos, social networking details, and other data.

Dominic's plans for lunch—and for most of the coming days—were abandoned. He did an about-face and bolted up the stairs, two at a time, back to the fifth floor. He suspected this issue was connected to a troubling piece published that morning by the *Guardian*. The newspaper had published a secret court order requiring American telecom giant Verizon "on an ongoing, daily basis" to hand over to government authorities its records of calls made both within and between the United States and other countries.[1] The records were analyzed by the NSA, which was headquartered in Fort Meade, Maryland, and had long collected signals intelligence and data around the world. According to the article, this bulk collection also targeted millions of Americans, regardless of whether they had done anything wrong.

If anyone at Microsoft would know about PRISM, it was John Frank, the lawyer who ran the legal teams that included our national security work. Dominic made a beeline for John's office.

Always measured and methodical, John slowly digested the *Guardian*'s message on Dominic's phone. He removed his glasses, leaned away from his desk, and gazed out into the sun-dappled day. He suddenly looked tired. "This makes absolutely no sense. None of this sounds right."

John not only knew how and what the company reviewed and responded to from law enforcement, he had helped design the process. Microsoft disclosed customer data only in response to valid legal process—and only for specific accounts or individuals.

When John and Dominic arrived at my office door, they had little more to share than the reporter's message. "If they're doing this, it's without our knowledge," John said.

Yes, we were obligated to review and respond to requests for user data in accordance with the law. We had an established process for carefully reviewing and responding to all data requests from law enforcement. But Microsoft is a huge company. Could this be the work of a rogue employee?

It was a question that we soon dismissed. We knew our engineering systems and the process for receiving, reviewing, and responding to government demands. The *Guardian*'s note just didn't add up.

No one at Microsoft had heard of PRISM. The *Guardian* was unwilling to disclose the leaked documents they were working from.

We reached out to people we knew at the White House, and they too would not speak of or share anything "classified." As the afternoon passed, I mused to John and Dominic, "Perhaps we're part of a secret club that is so secret we don't even know we're a member."

We'd have to wait to until after the story was published to even begin to respond to the reporter.

At three p.m. Pacific Daylight Time, the *Guardian* launched its bombshell: "NSA Prism Program Taps in to User Data of Apple, Google and Others."[2] We finally learned that PRISM, the NSA's national security electronic surveillance program, is an acronym for Planning Tool for Resource Integration, Synchronization, and Management.[3] Who had dreamed up that mouthful of a name? It sounded like a bad product name from the tech sector. According to the news media, it was an electronic surveillance program to track mobile devices, calls, emails, online conversations, photos, and videos.[4]

Within hours, the *Guardian*'s article and similar reporting by the *Washington Post* reverberated around the globe. Our sales teams and lawyers were inundated with customer calls.

They all had the same question: Was it true?

At first it was unclear where the media was getting its information. People debated whether it was even legitimate. But three days later, the newspaper dropped almost as big a bombshell as the reports themselves. The *Guardian* revealed its source,[5] at his own request.

That source was a twenty-nine-year-old employee of defense contractor Booz Allen Hamilton. His name is Edward Snowden. He worked at the NSA's Threat Operations Center in Hawaii as a contract computer systems administrator. He had downloaded more than a million highly classified documents,[6] and on May 20, 2013, he boarded a flight to Hong Kong, where he connected with journalists from the *Guardian* and the *Washington Post* and began to share the NSA's secrets with the world.[7]

Snowden's documents would turn into a series of news stories that summer and fall. The first document leaked was a classified forty-one-slide PowerPoint presentation used to train intelligence personnel. But that was just the start. Reporters would milk Snowden's stash of secret files into the next year, stoking anxiety with a steady wave of headlines. A tsunami of public mistrust built as claims surfaced of the US and UK governments accessing phone records and user data, including information belonging to foreign leaders and millions of innocent Americans.[8]

The news struck a nerve with the public, and for good reason. The assertions flew in the face of the privacy protections that democratic societies had taken for granted for more than two centuries. These rights, which we rely on to protect your information in our Quincy data center today, were born in the eighteenth century during a boiling controversy in the streets of London. The man who ignited the political firestorm was a member of Parliament himself. His name was John Wilkes.

John Wilkes was arguably the most dramatic—and radical—politician of his time. In the 1760s, he challenged not only the Prime Minister but also the King with words so colorful they would make some of today's politicians blush (almost). In April 1763, Wilkes penned an anonymous critique in an opposing periodical. The article infuriated British attorney general Charles Yorke, who suspected the author was Wilkes, and soon the government issued a search warrant that was so broad, the officers of the peace had the authority to search almost any place at any time.

Acting on the flimsiest information, they entered the house of a suspected printer in the middle of the night, "took him out of his bed from his wife, seized all his personal papers, and arrested fourteen journeymen and servants."[9] The British authorities quickly followed by searching four more houses, arresting a total of forty-nine people, almost all of whom were innocent. They knocked down doors, ransacked trunks, and broke hundreds of locks.[10] They eventually had enough evidence to nab the man they wanted; John Wilkes was placed under arrest.

Wilkes was not the type to take things lying down. Within a month, he'd filed a dozen lawsuits and had gone to court to challenge the most powerful officials in the land. While that was hardly surprising, what happened next shocked the British establishment and especially the government itself: The courts ruled in Wilkes's favor. Upending literally centuries of power exercised by the King and his men, the courts required that the authorities have greater probable cause to support a search, and even then, that they do so in a more limited manner. The British press hailed the rulings, citing the famous phrase that every Englishman's "house is his castle, and is not liable to be searched, nor his papers pried into, by the malignant curiosity of the King's messengers."[11]

In important respects, John Wilkes's lawsuits marked the birth of modern-day privacy rights. These rights were the envy of free people, including the British colonists living in North America. Just two years earlier, they had pursued—and lost—an equally hot dispute in New England, where even before becoming a barrister, John Adams, then in his midtwenties, sat in the back of a Boston courtroom to watch one of the continent's greatest showdowns of the early 1760s. James Otis Jr., one of the most fiery lawyers in Massachusetts, had protested that British troops were using powers much like those that Wilkes contested. As local merchants smuggled imports without paying taxes they regarded as unjust, the British had responded by using so-called general warrants to go from house to house looking for customs violations without specific evidence.[12]

Otis argued that this was a fundamental violation of civil liberties, calling it "the worst instance of arbitrary power."[13] While Otis lost his case, his words marked the colonists' first step toward rebellion. Near the end of his life, Adams would still remember Otis's argument and write that it had "breathed into this nation the breath of life."[14] He would say until the day he died that it was that day, that case, that courtroom, and that issue that set the United States on a course toward independence.[15]

It would take thirteen years after the Declaration of Independence to realize the principle that Otis had advocated so passionately. By then the issue had moved to New York, where the first US Congress assembled on Wall Street in 1789. James Madison stood before the House of Representatives and introduced his proposed Bill of Rights.[16] It included what became the Fourth Amendment to the Constitution, guaranteeing that Americans would be secure in their "persons, houses, papers, and effects" from "unreasonable searches and seizures" by the government, including the use of general warrants.[17] Ever since, the authorities have been required to go to an independent judge and show "probable cause" to obtain a warrant to search a home or office. In effect, this means the government must demonstrate to a judge that there exist facts that would lead "a person with reasonable intelligence" to believe that a crime is being committed.[18]

But does this protection extend to information that leaves your home? The Fourth Amendment was put to this test after Benjamin Franklin invented the post office. You seal an envelope and give it to an agency of the government itself. The Supreme Court in the 1800s had no trouble finding that people still had a right to privacy in their sealed letters.[19] As a result, the Fourth Amendment applies and the government cannot open an envelope and look inside without a search warrant based on probable cause, even though the government's postal service is in possession of that envelope.

Over the centuries, the courts looked at whether people had a "reasonable expectation of privacy" and considered what it meant when you stored your information with someone else. Put simply, if it was in something like a locked storage container and the key was inaccessible to others, then judges concluded that there was such an expectation and the Fourth Amendment applied. But if you stored your documents in a box of files that was stacked next to other people's boxes where people could come and go, then the police didn't need a search warrant. This was because the courts concluded that you had abandoned your reasonable expectation of privacy under the Fourth Amendment.[20]

Today's fortified data centers with redundant layers of physical and digital security seem to adequately qualify for a locked storage container.

In the summer of 2013, we were hit regularly by one reporter after another pursuing the Snowden story based on a newly leaked classified document. A routine developed. When I saw Dominic huddled in John's office, I knew another story was about to publish. Most times, we didn't even know what we were responding to. "I'd have the same conversation with a different reporter on an almost daily basis those first few weeks," recalled Dominic. "They'd say, 'Well Dominic, someone is lying. It's either Microsoft or Edward Snowden.'"

The *Guardian*'s reporting about PRISM reflected only one part of a longer story about the NSA's efforts to obtain data from the private sector. As declassified documents have now exhaustively detailed,[21] in the days after the tragedy on September 11, 2001, the agency pursued voluntary partnerships with the private sector to collect user data outside the legal subpoena and warrant process.

Microsoft, like other leading technology companies, wrestled with whether to provide such data to the government voluntarily. As we talked through these questions internally, we appreciated the broader geopolitical climate. The long shadow of the September 11 attacks hung over the nation. Coalition forces had unleashed Operation Enduring Freedom in Afghanistan, Congress was supporting the invasion of Iraq, and a fearful American public was calling for stronger antiterrorism efforts. It was an extraordinary time. As many said, it called for an unprecedented response.

But there was a fundamental problem with an approach that asked companies voluntarily to turn over information like what was described in the declassified reports. Data sought by the NSA didn't belong to technology companies. It belonged to customers, and it included some of their most private information.

Like the PRISM program, the NSA's post–September 11 efforts to obtain customer information voluntarily from the private sector raised a fundamental question: How can we fulfill our responsibility to customers while answering the call to protect the country?

To me, the answer is clear. The rule of law should govern this issue. The United States is a nation governed by laws. If the US government wants our customers' records, it needs to follow the law of the land and go to court to get them. And if the officials in the executive branch didn't think the law went far enough, they could go to Congress and ask for more authority. That's how a democratic republic should work.

While in 2002 we could not have predicted Edward Snowden and his famous flight, we could look back at history to predict more generally what was likely in store for the future. In times of a national crisis, trade-offs between individual freedoms and national security were nothing new.

The nation's first such crisis occurred a little more than a decade after the Constitution was signed. It was 1798 when a "quasi war" broke out between the United States and France on the Caribbean Sea. The French, wanting to pressure the United States to repay loans made by its overthrown monarch, seized more than three hundred

American merchant ships and demanded ransom.[22] Some angry Americans called for outright war. Others, such as President John Adams, thought the new nation was no match for the French. Fearing that public debate would fatally undermine the fledgling government, Adams sought to quell the discord by signing a set of four laws that became known as the Alien and Sedition Acts. These acts allowed the government to imprison and deport "dangerous" foreigners and made it a crime to criticize the government.[23]

Some sixty years later, during the Civil War, the United States would again set aside a key tenet of our democracy when President Abraham Lincoln suspended the writ of habeas corpus several times to suppress Confederate rebellions. To enforce the army's draft, Lincoln broadened the suspension and denied the right to trial nationwide. All in all, as many as fifteen thousand Americans were held in prison during the war without appearing before a judge.[24]

In 1942, shortly after the bombing of Pearl Harbor, President Franklin D. Roosevelt, swayed by the military and by public opinion, signed an executive order forcing 120,000 Americans of Japanese descent into remote camps, caged in by barbed wire and armed guards. Two-thirds of those imprisoned had been born in the United States. When the order was rescinded three years later, most had lost their homes, farms, businesses, and communities.[25]

While the country accepted those injustices in moments of national crisis, Americans later questioned the price they had paid for public safety. In my mind, one question was "How will we be judged ten years from now, when the moment passes? Will we be able to say that we honored our commitment to our customers?"

Once the question was apparent, the answer was clear. We can't turn over customers' data voluntarily without valid legal process. And as the company's most senior lawyer, I have to take responsibility—and bear any criticism—for this position. After all, who better than the lawyers to defend the rights of the customers we serve?

Against this backdrop, virtually every leading tech company found itself on the defensive in the summer of 2013. We conveyed our frustration to officials in Washington, DC. It was a watershed moment. It surfaced contrasts that have contributed to a chasm between governments and the tech sector to this day. Governments serve constituents who live in a defined geography, such as a state or nation. But tech has gone global, and we have customers virtually everywhere.

The cloud has not only changed where and to whom we provide our services, it has redefined our relationship with customers. It has turned tech companies into institutions that in some ways resemble banks. People deposit their money in banks, and they store their most personal information—emails, photos, documents, and text messages—with tech companies.

This new relationship also has implications beyond the tech sector itself. Just as public officials concluded in the 1930s that banks had become too important to the economy to be left unregulated, tech companies have become too important to be left to a laissez-faire policy approach today. They need to be subject to the rule of law and more active regulation. But unlike the banks of the 1930s, tech companies today operate globally, making the whole question of regulation more complicated.

As customer discontent grew around the world in 2013, we realized there was no way we could address their concerns without saying more. We knew well the clear limitations we had imposed on our own services and the sometimes complicated work to address the preexisting practices of companies that we later acquired. We wanted to explain that we only turned over customer information in response to search warrants, subpoenas, and national security orders. But when we proposed communicating this publicly, the Department of Justice, or DOJ, told us the information was classified and we could not. Frustration built.

We decided to do something that we had never done before: sue the United States government. For a company that had fended off a decade of government antitrust litigation and then spent another decade working to make peace, it felt like we were crossing a new Rubicon. We moved forward with a motion that initially was kept secret in the Foreign Intelligence Surveillance Court, or FISC.

The FISC is a special court established to review the government's surveillance orders. It was created during the Cold War to approve wiretaps, electronic data collection, and the monitoring of suspected terrorists and spies. It is shrouded in secrecy to protect intelligence efforts to monitor and thwart security threats. Each warrant issued under the Foreign Intelligence Surveillance Act comes with a gag order that prohibits us from telling our customer that we've received a warrant for their data. While this was understandable, our legal case asserted that we had a right to share broader information with the public under the Constitution's First Amendment and its commitment to freedom of expression. At a minimum, we argued, this gave us the right to talk generally about the number and types of orders we received.

Soon we learned that Google had done the same thing. This led to a second watershed moment. For five years our two companies had battled our differences before regulators around the world. Google argued for restrictions on Windows. Microsoft argued for restrictions on Google searches. We knew each other well. I had a lot of respect for Kent Walker, Google's general counsel. But no one would have accused us of being best friends.

Suddenly we were on the same side in a new and common battle with our own government. I decided to reach out to Kent, at first without luck as we traded messages. As I left an employee town hall on a July morning in one of the buildings where our Xbox team worked, I pulled out my cell phone to try again. I looked for a quiet corner and found myself standing next to a life-size cardboard cutout of Master Chief, the soldier who leads the troops in our *Halo* game into war against an alien enemy. I liked that Master Chief had my back.

Kent answered the phone. While we had talked many times before, it was almost always to discuss the complaints our companies had with each other. Now I proposed something different. "Let's join forces and see if we can negotiate with the DOJ together."

I would not have blamed Kent if he suspected a Trojan horse. But he listened and came back to me a day later saying he wanted to work together.

We held a joint call with the government to try to negotiate common terms. It seemed as if we were getting close to a settlement, when suddenly in late August the negotiations ended in failure. From our vantage point, it seemed as if the NSA and FBI were not on the same page. As summer faded into fall in 2013, Snowden's continued disclosures drove a deeper wedge between the US government and the tech sector. And then things went from bad to worse.

On October 30, the *Washington Post* published a story that set the industry's hair on fire: "NSA Infiltrates Links to Yahoo, Google Data Centers Worldwide, Snowden Documents Say."[26] The story was co-authored by Bart Gellman, a journalist I had known and respected since he wrote for the *Daily Princetonian* at Princeton University, where we were undergraduates together. His article said that the NSA, with the help of the British government, was surreptitiously tapping into undersea fiber-optic cables to copy data from Yahoo and Google networks. While we could not verify whether the NSA was targeting our cables, some of Snowden's documents also referred to our consumer email and messaging services.[27] That made us suspect we had been tapped as well. To this day, the US and British governments have not spoken publicly to deny hacking into data cables.

The tech sector responded with a combination of astonishment and anger. At one level, the story provided a missing link in our understanding of the Snowden documents. It suggested that the NSA had much more of our data than we had lawfully provided through national security orders and search warrants. If this was true, the government in effect was conducting a search and seizure of people's private information on a massive scale.

The *Washington Post* story indicated that the NSA, in collaboration with its British counterpart, was pulling data from the cables used by American technology companies, potentially without judicial review or oversight. We worried that this was happening where cables intersected in the United Kingdom. As lawyers across the industry compared notes, we theorized that the NSA persuaded itself that by working with or relying upon the British government and acting outside US borders, it was not subject to the Fourth Amendment to the US Constitution and its requirement that the NSA search and seize information only pursuant to due process and court orders.

The reaction at Microsoft and across the industry was swift. In the weeks that followed, we and other companies announced that we would implement strong encryption for all the data we moved between our data centers on fiber-optic cables, as well as for data stored on servers in our data centers themselves.[28] It was a fundamental step in protecting customers, because it meant that even if a government siphoned up customer data by tapping into a cable, it would almost certainly be unable to unlock and read what it had obtained.

These types of encryption advances were easier said than done. They would involve large computational workloads for our data centers and require substantial engineering work. Some of our engineering leaders were less than enthusiastic. Their concerns were understandable. Software development inherently involves choices between features, given the finite availability of engineering resources that can be applied on a feasible timeline. This encryption work required them to delay the development of other product features that customers were asking us to add. After some animated discussion, CEO Steve Ballmer and our senior leadership team made the decision to press forward quickly on the encryption front. Every other tech company did the same thing.

That November, as these events were unfolding, President Barack Obama visited Seattle. He was attending a political fund-raiser, and the White House had invited a small group of area leaders and supporters to have a cocktail in a private suite at the Westin Seattle hotel after the formal event. I was invited to represent Microsoft.

I hoped that this occasion would allow me a few minutes to talk with the president about the First Amendment issues we had raised in our lawsuit. But the Justice Department lawyers had asked us not to raise the case with him. "Their client" was represented by counsel, and all conversations had to go through them. But just before President Obama arrived in the room, I asked his assistant, Valerie Jarrett, whether it would be appropriate to ask him a different question that wasn't regarding our lawsuit: whether he thought the Fourth Amendment's protections against unreasonable searches and seizures by the government guarded Americans even outside the United States.

Given the *Washington Post*'s report about the NSA tapping cables run by American companies outside the United States, I thought it was an important question. Valerie thought he'd find the topic interesting.

She was right. As I spoke with the president, the former constitutional law professor emerged. While President Obama had clearly mastered more constitutional law than I remembered, I recalled enough to have a respectable conversation.

And then he changed the subject.

"I heard you all didn't want to settle your lawsuit with us. You think it's better if you're perceived as suing the government. Is that right?" It was one of those moments that called for an instant mental calculation. The Justice Department's lawyers had certainly never instructed us not to answer direct questions from the President of the United States, so answer I did, explaining that we'd wanted to settle but it seemed the government did not. I described our concerns and belief that we could make real progress if we could get the right people in the same room.

A few weeks later, Obama invited a group of tech leaders to the White House. It was eight days before Christmas, and the West Wing, dressed in its holiday best, was at full tilt as the staff hurried to wrap up their work before the president left for his annual vacation in Hawaii. The White House had announced publicly that the meeting would address "health, IT procurement, and surveillance issues." It was a bit like telling baseball fans that they could go to an event that included the national anthem, a hot dog eating contest, and the first game of the World Series. We all knew what brought us to Washington on that cold winter morning.

An all-star cast of tech leaders arrived at the West Wing, including Apple CEO Tim Cook, Google chairman Eric Schmidt, Facebook COO Sheryl Sandberg, Netflix CEO Reed Hastings, and a dozen others. Most of us already knew each other. Eight of our companies—virtually all competitors—had just come together to create a new coalition, called Reform Government Surveillance, to work together on precisely the issues we were there to discuss. After a round of enthusiastic greetings, we put our smartphones in a rack of cubbies in the hallway and filed into the Roosevelt Room.

The Roosevelt Room is named not for one president, but for two—Theodore Roosevelt, who built the West Wing, and Franklin Roosevelt, who enlarged it. [29] As I took my seat at the long, polished conference table, I gazed up at a painting that crowned the fireplace and chuckled. It was of Teddy, the Rough Rider, atop his rowdy horse. Hopefully, the next ninety minutes wouldn't be as rough.

We were greeted by a White House that had similarly turned out in force. President Obama and Vice President Joe Biden took their customary seats at the middle of the table and were flanked by virtually the entire senior staff. The press corps took photos while the president asked Reed some safe questions about the upcoming season of *House of Cards*.

After the press left the room, the conversation took a serious tack. The custom at these meetings during the Obama administration was for each guest to offer some initial comments. With a group this large, that took a while. The president put his Socratic skills to work by asking questions and turning the recitation of talking points into a more penetrating conversation.

With only a couple of exceptions, each tech leader made a strong argument for restricting mass data collection, creating more transparency, and imposing more checks and balances on the NSA. For the most part, we steered clear of talking about Edward Snowden directly. But as the conversation worked its way around the table, Mark Pincus, the founder of the social game company Zynga, who was sitting near Obama, argued that Snowden had been a hero. "You should pardon him," Pincus said, "and give him a ticker tape parade."[30]

As Biden visibly recoiled, Obama said, "That's one thing I'm not going to do." The president explained that he thought Snowden had acted irresponsibly by taking so many documents and leaving the country.

Then it was Yahoo CEO Marissa Mayer's turn to speak. Sitting next to Pincus, she opened a manila folder with her carefully prepared talking points. She started by saying, "I agree with what everyone else has said," then paused and looked up. She quickly pointed to Pincus and added, "Except him. I don't agree with him." Everyone laughed.

The exchange reflected the needle we were all trying to thread. Almost all of us had turned up at the White House to press the president to change the government's course. But the tech sector had a cordial and even warm relationship with Obama, and it's always harder to challenge someone when you're visiting his or her house. Especially when it's the White House.

While we were all polite, we stuck to our guns and made the case for surveillance reform. It was clear that Obama had given the whole topic considerable thought, describing the list of issues that he thought the government needed to address. He sometimes pushed back, saying that while people were concerned about all the data possessed by the NSA, the companies around the table collectively had far more data than the government. "I have a suspicion that the guns will turn," he said.

At the end of the meeting, the president made clear that he was interested in pursuing several important if limited changes in US policy. He reeled off a subset of issues and asked people to provide more information to help take the conversation "to the next level of detail."

A month later, on January 17, 2014, the president took the first important steps toward surveillance reform.[31] The night before he unveiled his plans, we got a call from the DOJ's lawyers. They offered to settle the cases brought by Microsoft and Google on terms even more favorable to our position than those we had said we would accept in our negotiations the preceding August. Once the settlement was in place, our companies moved forward with new transparency reports to publish more data about national security warrants and orders, with Google, to its credit, getting off to the fastest start with an impressive model that the rest of us decided to follow.

For many customers and privacy advocates, Obama's speech represented a first step, with many more needed. Across the tech sector, we endorsed these views. We recognized that the issues were not easy and that difficult questions remained. How could we reassure foreign governments and customers that the US government would not reach inappropriately into the data centers run by US companies? How could we simultaneously take the lawful steps needed to keep the public safe? These would take years to resolve.

It was remarkable to consider how much had changed since Snowden handed over his stolen documents to the *Guardian* seven months earlier. People's eyes were opened to the scope of government surveillance. Stronger encryption had become the new norm. Tech companies were suing their own government. And competitors were working together in new ways.

Years later, people still debate whether Edward Snowden was a hero or traitor. In the eyes of some, he was both. But by early 2014, two things were clear: He had changed the world; and across the tech sector, he had changed us as well.

Chapter 2

⋙ TECHNOLOGY AND PUBLIC
SAFETY: "I'd Rather Be a
Loser Than a Liar"

The public depends on law enforcement to keep it safe. But you can't catch criminals or terrorists if you can't find them—and this requires effective access to information. In the twenty-first century, that information often resides in the data centers of the world's largest tech companies.

As the tech sector tries to do its part to keep the public safe and protect people's privacy, we've found ourselves perched atop a razor's edge. It's a delicate balance that we must maintain while we respond to a fluid and fast-changing world.

Events requiring our response arise suddenly and without warning. It's a reality I first grappled with in 2002. On January 23 of that year, *Wall Street Journal* reporter Daniel Pearl was abducted in Karachi, Pakistan.[1] His kidnappers moved between internet cafés using our Hotmail email service to communicate their ransom demands, kicking off a desperate manhunt by the Pakistani police. In exchange for Pearl, the kidnappers demanded the release of terrorist suspects in

Pakistan and the halt of a planned shipment of F-16 fighters from the United States. It was clear the Pakistani government would not agree to the ransom demands. The only way to save Pearl was to find him.

The Pakistani authorities worked quickly behind the scenes with the FBI in the United States, who came to us. Congress had created an emergency exception to the Electronic Communications Privacy Act so that the government could act immediately and tech companies could move quickly when there was an "emergency involving the danger of death or serious physical injury."[2] Pearl's life was clearly in jeopardy.

John Frank came to me and explained the situation. I gave the green light to work with the local police and the FBI. Our goal was to monitor the Hotmail account being used by the kidnappers and use the IP address in their newly created emails to locate the internet cafés halfway around the world where they were sitting. Our teams worked closely with the FBI and the local authorities in Pakistan for a week, trailing the kidnappers as they bounced from hotspot to hotspot accessing the internet.

We came close but not close enough. The kidnappers killed Pearl before being caught themselves. We were devastated. His brutal death underscored the enormous stakes and responsibility that had been cast upon us, something we seldom spoke about publicly.

The incident was an early indicator of what was to come. Today, cyberspace is no longer some peripheral dimension. It increasingly has become the place where people organize themselves and define what happens in the real world.

The tragedy involving Daniel Pearl also underscores the importance of exercising judgment in terms of privacy. In important ways, there is a balance between privacy and safety that benefits from privacy groups that push in one direction and law enforcement agencies that push in another. But like the judges who decide these disputes, tech companies have become a place where these issues come to a head. We need to understand and think hard about both sides of this equation.

One big challenge is how to do this well. Our ability to turn on a dime in response to search warrants is a process that was honed through trial and error since the birth of email and electronic documents in the 1980s.

In 1986, President Ronald Reagan signed the Electronic Communications Privacy Act, affectionately known by today's privacy lawyers as ECPA. At the time, no one knew whether the Fourth Amendment would protect something like electronic mail, but Republicans and Democrats alike wanted to create this type of statutory protection.

As sometimes happens in Washington, DC, in 1986 Congress acted with good intentions but in a way that was far from simple. Part of ECPA was the Stored Communications Act, which created what was basically a new form of search warrant. With probable cause, the government could go to a judge, secure a search warrant for your email, and serve the warrant not on you but on the tech company where your email and electronic documents were stored.[3] The company was then obligated to pull the email and turn it over. In certain circumstances, the law in effect turned tech companies into agents of the government.

This also created a new dynamic. If the government served a search warrant at your home or office, someone was likely to be there and know what was happening. They couldn't stop it, but they were aware. If they thought their rights were violated, they could follow in John Wilkes's footsteps and go to court.

But Congress adopted a more complicated approach when it came to notifying people and businesses that the government was obtaining their emails and documents from tech companies, creating a statute that gave the government the authority to seek a gag order that would compel a tech company to keep the warrant secret. This statute gave the government five different bases on which to demand secrecy. On the surface, these bases did not look unreasonable. For example, if disclosure would lead to destruction of evidence or intimidation of a witness, or would otherwise jeopardize an investigation, a judge could issue a warrant together with what is called a nondisclosure order.[4] A tech company might receive both orders together, the first requiring it to turn over electronic data files and the second requiring it to keep the demand secret.

When email was still a rarity, these new warrants and gag orders were few and far between. But once the internet exploded and data center campuses emerged with hundreds of thousands of computers, life became far more complex. Today, twenty-five full-time employees—compliance experts, lawyers, engineers, and security professionals—make up our Law Enforcement and National Security team. They work with broad support provided by numerous law firms around the world, and they're known across Microsoft as the LENS team. Their mission is straightforward: to review and respond globally to law enforcement requests under the laws of different countries and in accordance with our contractual obligations to our customers. This is no small task. The LENS team operates from seven locations in six countries on three continents. During a typical year, they address more than fifty thousand warrants and subpoenas from more than seventy-five countries.[5] Only 3 percent of these demands are for content. In most cases, authorities are looking for IP addresses, contact lists, and user registration data.

When Microsoft receives a warrant, it typically comes through email. A compliance manager reviews the demand to ensure that it's valid and signed by a judge, that the authorities have probable cause, and that the agency has jurisdiction over the information. If everything checks out, the compliance manager will pull the requested evidence from our data center. The data is reviewed for a second time to make certain that we are only producing exactly what's specified in the warrant, and it's then sent to the requesting authority. As one LENS employee explained to me, "It sounds simple, but it takes a lot of time to do a good job. You need to review the warrant itself, review the account information associated with it, pull the information, and then review it again to be certain that what you're providing is appropriate."

When a compliance manager concludes that a warrant is too broad or the request exceeds an agency's jurisdiction, the case is escalated to an attorney. Sometimes we ask for warrants to be narrowed. Other times we deem the warrant unlawful and refuse to comply.

One member of the LENS team is on call 24-7, meaning for a week at a time he or she will sleep next to a phone in case there is an emergency or terrorist incident somewhere in the world that requires immediate action. During weeks when the world is turned upside down, members of the LENS team take turns being on call, so each person gets enough sleep to be alert on the job.

In 2013, as Edward Snowden shared the NSA's secrets and the public issues relating to this massive amount of data started to explode, a new lawyer joined Microsoft to lead the team. Her name is Amy Hogan-Burney. Armed with a keen intellect and sharp sense of humor, she quickly won over the team. Amy had spent the prior three years as an attorney in the National Security division at the FBI's headquarters. It equipped her well for the work at Microsoft, even if there would be days when she was on the opposite side of an issue from her former colleagues in Washington, DC.

Amy quickly adapted to her new role. She sat just downstairs from my office and I found myself walking down to her corridor more and more frequently. Her office was next to Nate Jones's, who had joined Microsoft earlier in the year after wrapping up more than a decade serving in the US government, including time with the Senate Judiciary Committee, the Department of Justice, and finally on President Obama's National Security Council working on counterterrorism.

Amy managed the work of the LENS team while Nate managed our overall compliance strategy, our relations with other tech companies, and negotiations with international governments. As the world had evolved, they and the entire LENS team had to strike a delicate balance. They needed to work with law enforcement agencies around the world, but they were also on the front line defending the privacy rights enshrined in the Fourth Amendment and other countries' laws. As they worked with the multiple privacy experts we already had on board, I was glad their offices were close to mine.

Nate and Amy quickly became something of a tag team, so much so that others on the team began referring to them as "Namy." Across Microsoft, people relied on Nate and Amy to work together quickly to think through our approach on the most sensitive issues. Our compliance managers would glance at a hot issue that arrived in their inbox, talk with each other, and decide that it needed to go to Namy right away.

Our Namy team was in the hot seat for protecting the world's filing cabinet—a seat that often got hotter in sudden and dramatic ways.

As office workers across France prepared to break for lunch on Wednesday, January 7, 2015, two brothers entered the Paris headquarters of the satirical magazine *Charlie Hebdo* and viciously murdered twelve people.[6] The two men were affiliated with Al-Qaeda, and they had been offended, as had many other Muslims, by the publication's profane cartoons of the prophet Muhammad.[7] But unlike many others, these brothers had taken matters into their own hands.

The tragedy was all over the news. We saw the horrific events unfold from Redmond with the rest of the world. As I refilled my coffee mug in the break room, a group of us watched the television as French police searched for the two brothers, who had managed to escape. Soon soldiers from the French army were involved in a nationwide manhunt, and another Al-Qaeda member launched a separate deadly terrorist attack in a French supermarket.[8] I recognized the streets and neighborhoods involved; I'd spent my first three years as a Microsoft employee working at our European headquarters in Paris.

Other than checking on our employees in the area, who were all safe, the story seemed important for the world but unrelated to my own job. That was no longer the case when the sun rose the next day in Redmond. France's national police quickly determined that the two terrorists had Microsoft email accounts, and they asked the FBI for help. At 5:42 a.m. in Redmond, the FBI in New York responded to the emergency and sought from us the killers' email and account records, including the IP addresses that can show the location of a computer or phone when a user logs in. A team at Microsoft reviewed the emergency request and provided the information to the FBI within forty-five minutes. A day later, the national French manhunt led authorities to the two terrorists, who were killed in a shootout with police.

The events in Paris shook France and the world. The Sunday following the attack, more than two million people marched in the French capital's streets to mourn the journalists and stand in solidarity, demonstrating support for freedom of the press.[9]

Unfortunately, it was not the last tragedy inflicted upon Paris in 2015. On a Friday evening in November, as Parisians were winding down from the workweek, terrorists struck again in coordinated attacks across the city. They opened fire with automatic rifles at a concert inside a theater, outside a stadium, and in restaurants and cafés.

The scenes were horrifying. The terrorists killed 130 people and injured more than five hundred others. It was the deadliest attack Paris had experienced since World War II. And while seven of the attackers were killed, two others managed to escape.[10]

French President François Hollande immediately declared a state of emergency across the country. The Islamic State of Iraq and Syria—ISIS—claimed responsibility, and it soon became apparent that some of the attackers had come from Belgium. A new manhunt ensued, this time spanning two nations.

Working with European authorities, the FBI again quickly served warrants and subpoenas on tech companies for the email and other accounts belonging to the suspects. We had learned from the *Charlie Hebdo* tragedy that we needed to be ready to spring into action when terrorists struck. This time, the authorities in France and Belgium served fourteen orders on us. The team reviewed them, determined they were lawful, and provided the information requested, in each case turning around information in less than fifteen minutes.

The two tragedies in Paris were events that grabbed the world's attention. But the days they occurred were far from the only days that demanded our own. When email was in its infancy, governments rarely turned to us. But once fifty thousand search warrants and government orders started arriving from more than seventy countries each year, we needed to operationalize our work on a global scale.

Satya Nadella helped define our path forward. He'd run Microsoft's cloud business before becoming the company's CEO in early 2014. More than anyone else, he understood the cloud. He also brought another valuable sensibility to this complex issue. He had grown up as the son of a senior civil servant in India, where his father was revered as the leader of the academy that trained a generation of the country's senior civil servants in the decades following the nation's independence. This background gave Satya an intuitive feel for how governments worked. I was struck by the similarity to Bill Gates, who had grown up as the son of one of Seattle's most prominent and respected lawyers. Bill and Satya were both quintessential engineers, but Bill could think like a lawyer and Satya could think like someone in government. For me, the opportunity to talk through tough issues with both was invaluable.

As we grappled with the full range of surveillance issues, Satya suggested in late 2014 that we needed to develop a principled approach. "We need to know how to make the hard calls, and we need our customers to know how we're doing it," he said. "And we need a set of principles to guide this work."

We had applied a similar approach to hard issues over the preceding decade, including the publication of "Windows Principles: Twelve Tenets to Promote Competition" to address our antitrust issues. I had unveiled those principles at the National Press Club in Washington, DC, in 2006.[11] Jon Leibowitz, then a member of the Federal Trade Commission who had pushed us on the subject amid our high-profile antitrust cases, attended the speech and came up to me afterward. "If you had come out with these a decade ago," he said, "I don't think the government would have sued you."

While Satya's assignment seemed straightforward, it wasn't. We needed principles that could apply across our entire business from our operating systems to the Xbox. These principles had to be simple and memorable—not twenty paragraphs filled with legal and technical jargon. Coming up with something that is shorter and simpler is always more difficult.[12]

While the issue was complex, the starting point was not. We were always clear in our minds that the information people stored in our data centers didn't belong to us. People still owned their emails, photos, documents, and instant messages. We were stewards of other people's possessions, not the owners of this data ourselves. And as good stewards, we needed to use this data in ways that served its owners, rather than thinking about only ourselves.

Building from this starting point, we assembled a team that developed what would become four principles that we would call our "cloud commitments": privacy, security, compliance, and transparency. I loved pointing out to the company's marketing leaders that the lawyers had found a way to take a complicated topic and reduce it to four words. Not surprisingly, they were quick to point out that this was a first.

Still, developing clear principles and putting them to work were two separate challenges. The team built out each principle with details and training. The real test would come when new circumstances raised tough questions and required us to decide how far we would go to stand up for the commitments we had created.

One of the toughest calls soon came regarding our commitment to transparency. We recognized that transparency was a linchpin for everything else. If people didn't understand what we were doing, they could never trust us on anything else.

Our business customers, in particular, wanted to know when we received a warrant or subpoena seeking their emails or other data. We believed there was seldom a good reason for the government to serve legal orders on us rather than our enterprise customers. Unlike individual criminals or terrorist suspects, a reputable company or business was far less likely to run for the border or act unlawfully to thwart an investigation. And if the government was concerned about the potential erasure of data, we could act under a limited "freeze order" to make a copy of a customer's data while the government hammered out the legal issues with the customer before obtaining access to it.

In 2013, we stated publicly that we would notify our business and government customers if we received legal orders for their data.[13] If a gag order prohibited us from telling them, we'd challenge the order in court. We'd also direct government agencies to go straight to our customers for information or data about one of their employees—just as they did before these customers moved to the cloud. And we'd go to court to make it stick.

We faced our first test when the FBI served us with a national security letter seeking data that belonged to an enterprise customer. The letter barred us from telling the customer that the FBI wanted its data. We studied the letter and could see no reasonable basis for the FBI to prohibit us from notifying the customer, let alone demand the data from us rather than obtain it directly from the customer. We refused, filed a lawsuit, and went to federal court in Seattle, where the judge was sympathetic to our argument. The FBI got the message and withdrew the letter.

Over the next year, our lawyers made good progress in pushing the Justice Department to go directly to enterprise customers for data. But in January 2016, an assistant US attorney in another district disagreed and served a search warrant under seal on us demanding data that belonged to a business customer. He coupled the warrant with an open-ended gag order that would stay in place forever. We objected.

Typically, once we had explained our position, the government would back off. This time, the federal attorney persisted and forced us to go to court.

I was traveling in Europe and woke early to an email from David Howard, who was responsible for our litigation work and several other areas. David had joined us five years earlier and was a successful former federal prosecutor and law firm partner. He brought a calm demeanor and good judgment to every tough problem. His leadership had been instrumental in what became year after year a pattern of winning 90 percent of our lawsuits. As I said only half-jokingly to our board of directors, "I've learned from David that good litigation results are not actually hard to achieve. You just have to fight the cases you deserve to win and settle the cases you deserve to lose." The key was having someone like David who could discern the difference.

In this instance, David wasn't optimistic about our chances. The judge was not sympathetic and was threatening to hold us in contempt of court. David wrote that the litigation team wanted to turn over the customer's data to avoid a fine.

On a conference call later that day, I told the team that I didn't want to surrender. We had made a promise to our customers to fight these types of orders, and that included going to court and taking on tough battles.

One of the litigators said that this was clearly a fight we would lose, and it could be an expensive defeat. "I'd rather be a loser than a liar," I said. "A promise is a promise." I felt that the cost of breaking it was greater than any amount of money, even if the outcome remained under seal and kept a secret.

I told the litigation team that if they fought the case, lost the battle, and kept the fine under $20 million, I would consider it a moral victory. We all knew there was no way we'd receive a fine for more than a fraction of that amount. It was my way of telling the litigators—who thankfully wanted to win every case—that there was no way they could lose this one as far as I was concerned.

The Microsoft team worked around the clock and through the weekend with our outside lawyers. We lost the case, but we avoided the contempt fine entirely and preserved the ability to be transparent with our customers and state generally that we had now lost one of these cases. And most important, we had lived up to our promise.

We worried that we'd continue to be tested in this way on a case-by-case basis. We needed to go on the offensive. "We aren't going to win these cases if we let the government pick every fight," David said. "These types of gag orders are supposed to be the exception, not the norm. But the government is making them routine. We need the courts to rule on this broad practice."

He developed a brilliant play. We decided to pursue what is called a declaratory judgment, which would clarify our rights. We argued that the government was exceeding its constitutional powers by routinely issuing gag orders under the Electronic Communications Privacy Act. We combed through available warrant records from the prior year and a half and found that more than half of government data demands for individuals were bound by gag orders, with half of these written to ensure they were kept secret forever.

We returned to federal court in Seattle to sue our own government. We argued that the excessive use of gag orders violated our First Amendment right to tell our customers that the government was seizing their emails. We also maintained that these gag orders violated our customers' Fourth Amendment right to be protected from unlawful search and seizure, because people had no way of knowing what was happening and weren't able to stand up for their legal rights.

The case squarely raised whether people's rights would be protected in the cloud. We were optimistic, bolstered by the trend we saw unfolding in the Supreme Court.

In 2012, Supreme Court justices declared in a 5–4 decision that the Fourth Amendment required that the police get a search warrant before putting a GPS locator on a suspect's car.[14] While the other justices found that the "physical intrusion" of attaching a device to someone's car required a search warrant, Justice Sonia Sotomayor recognized that in the twenty-first century, law enforcement didn't necessarily need to physically intrude to track someone's location. GPS-enabled smartphones, which create remote records of someone's location, were starting to spread. They revealed all sorts of personal information that the government could mine for years. As Sotomayor put it, unless this type of surveillance was safeguarded under the Fourth Amendment, it could "alter the relationship between citizen and government in a way that is inimical to democratic society."[15]

Justice Sotomayor captured something else that we thought was fundamental. For almost two centuries the Supreme Court had said the Fourth Amendment failed to protect information that was widely shared, on the theory that people no longer had a "reasonable expectation of privacy." Now, however, Sotomayor noted, privacy meant the ability to share information but determine who can see this information and how it will be used. She was the first justice to articulate this shift, and the big question was whether the other justices would embrace it.

Two years later, an answer began to emerge. In the summer of 2014, Chief Justice John Roberts wrote an opinion for a unanimous Supreme Court.[16] The justices decided that the police needed a warrant to search someone's cell phone, even if the person was under arrest for committing a crime. As Roberts put it, "Modern cell phones are not just another technological convenience. With all they contain and all they may reveal, they hold for many Americans the privacies of life."

While the Fourth Amendment was adopted to protect people in their homes, Roberts explained that modern phones "typically expose to the government far more than the most exhaustive search of a house: A phone not only contains in digital form many sensitive records previously found in the home; it also contains a broad array of private information never found in a home in any form."[17] Hence the Fourth Amendment applied.

We cheered when we read what Roberts wrote next. For the first time, the Supreme Court in effect addressed the files stored in our data centers, like the one in Quincy. "The data a user views on many modern cell phones may not in fact be stored on the device itself," he wrote. "The same type of data may be stored locally on the device for one user and in the cloud for another."[18] For the first time, the Supreme Court recognized that a search of a phone reached far beyond what was in a person's physical possession. In effect, new technology had created new grounds for strong privacy protection in the cloud itself.

While these words didn't speak directly to our protest of broad gag orders in Seattle, they provided some helpful tailwinds for our broader privacy cause. Now we needed to ride them.

We put David's plan into action by filing a lawsuit on April 14, 2016.[19] It was assigned to Judge James Robart, who had been a leading light in Seattle's legal community before becoming a federal judge in 2004. We had appeared before him previously, including in a big patent trial. He was tough, but smart and fair. He kept our litigators on their toes, and from my vantage point, that was a good place for them to be.

As we filed our lawsuit, we shared our data from the preceding eighteen months, which showed that we had received more than twenty-five hundred gag orders applicable to individuals, effectively silencing us from speaking to customers about the legal process seeking their personal information.[20] Notably and even surprisingly, 68 percent of the total contained no fixed end date at all. This meant that we effectively were prohibited forever from telling our customers that the government had obtained their data.

We recognized that we needed to couple our concerns about the DOJ's current practice with a blueprint for a better approach. We called for greater transparency and what we termed digital neutrality, or the recognition that people's information should be protected regardless of where and how it was stored. This should be balanced, we said, with a principle around necessity, so that gag orders could be issued but could be adapted to what's necessary for an investigation, and no more.

The government hit back with a motion to dismiss our lawsuit before it even got started. It argued that we had no right to inform customers under the First Amendment and no basis to stand up for customers' rights under the Fourth Amendment. We soon concluded that our ability to survive this motion would likely provide the critical turning point. If we survived, we would get access to government data on the broad use of secrecy orders, and this likely would give us the remaining facts we needed to drive our arguments across the finish line.

We decided that we needed to build a broad coalition of supporters. We spent the summer on a recruiting campaign. By Labor Day, more than eighty supporters had joined the case by filing amicus, or friend of the court, briefs. The group represented every part of the tech sector, the business community, the press, and even respected former officials from the Justice Department and the FBI.[21]

The lawyers and the public filed into Judge Robart's courtroom on January 23, 2017. It had been a year and two days since our decision to fight a gag order under seal rather than surrender. Now we had the opportunity for a public hearing on the government's motion, with former DOJ officials supporting us in the front row.

Two weeks later, Robart ruled that our case could proceed.[22] While he accepted the government's argument that we couldn't defend our customers' Fourth Amendment rights, he agreed that we had a basis to move forward with our First Amendment claim. We had lived to fight another day.

The Justice Department took notice and began to take our claims more seriously. We sat down and, after a number of discussions, the DOJ released a new policy that set clear limits on when prosecutors could seek gag orders. The department coupled this with new guidance directing prosecutors about going to enterprises before cloud providers in the case of enterprise warrants. We were satisfied, saying publicly that we thought the new approach would help ensure that secrecy orders are used only when necessary and for defined periods of time.[23] Both sides agreed to bring the gag order litigation to an end.

The outcome underscored the delicate balance between privacy and safety. Lawsuits typically are blunt instruments. By themselves, they can only determine if current processes are lawful. They can't craft a new proposal that addresses how technology should be governed. That requires real conversation and sometimes negotiation and even new legislation. In this case, the lawsuit had done what was needed, bringing everyone to the table to talk about the future. But getting everyone to sit down together on other issues remained an ongoing challenge, one that would become even more difficult and important.

Chapter 3 ⟫ # PRIVACY: A Fundamental Human Right

I n the winter of 2018, after a long day of public events and back-to-back meetings in Berlin, we were ready to call it a day. But Dirk Bornemann and Tanja Boehm, from our local German team, had a different idea. They insisted on one final stop, a former prison in the northeastern section of the city.

A week earlier, the opportunity for this diversion had piqued our curiosity, but the icy weather and jet lag had dampened our enthusiasm. This detour, however, turned out to be one of the most memorable days of the year.

The wintry light faded as we drove through the streets of the German capital. Through the car window a fast-moving reel of architecture told a tale of the city's past. Edifices dating back to Prussia, the German Empire, Weimar, and Nazi eras gave way to sterile Communist-era concrete blocks as we closed in on our destination: the former German Democratic Republic's Hohenschönhausen prison.

The once top-secret military compound had been part of the headquarters of the Stasi, short for State Security Service. The Stasi served as East Germany's "shield and sword," ruling over the country with repressive surveillance and psychological manipulation. By the time the Berlin Wall fell, the Stasi employed almost ninety thousand operatives backed by a secret network of more than six hundred thousand "citizen watchdogs" who spied on their East German coworkers, neighbors, and sometimes their own family.[1] The Stasi accumulated a staggering number of records, documents, images, and video and audio recordings that if lined up would stretch sixty-nine miles.[2] Citizens who were considered flight risks, threats to the regime, or asocial were detained, intimidated, and interrogated at Hohenschönhausen from the end of World War II until the end of the Cold War.

As the gate of the former prison swung open, we pulled past a concrete watchtower, where we were met by a seventy-five-year-old former prisoner, Hans-Jochen Scheidler. His athletic physique and easy smile belied his age and the ordeal he'd suffered at the prison. He shook our hands enthusiastically and led us into the large gray building where he'd spent a dark seven months.

In 1968, Scheidler left Berlin to pursue a PhD in physics at Charles University in Prague. "The Prague Spring was one of the happiest times of my life," he said, recalling the loosening of restrictions and political liberalization that took place in the capital that year. "Every weekend I celebrated Prague Spring there."[3] But Czechoslovakia's move toward liberty came to a swift end when half a million Warsaw Pact troops rolled into the country and suppressed the reforms.

That August, the twenty-four-year-old was home in Berlin when he heard the devastating news. The dream of a new era, one he considered a "more human version" of socialism, had been snatched away. In protest, Scheidler and four of his friends printed little leaflets criticizing the Soviet regime and slipped them into the mailboxes of East Berliners that night.

Caught in action, they were all arrested later that evening by the Stasi and sent to the very place where we now stood. He would spend seven months in one of the small, dark cells we visited, barred from seeing other inmates, talking with other people, or even reading a single piece of paper. His parents had no idea where he was or why he had vanished. He was subjected to cruel psychological torture. Even after his release, Scheidler wasn't allowed to study or work in his chosen field of physics.

The point of our visit that day was suddenly clear.

Today, much of the world's political activism doesn't start on the streets, as in Scheidler's time; it starts on the internet. Electronic communication and social media have provided a platform for people to mobilize support, spread messages, and voice dissent—accomplishing in days what would have taken weeks during the Prague Spring. Hans-Jochen had engaged in the 1960s equivalent of sending an email. And he was arrested while pushing "send."

When we talked about privacy issues inside Microsoft, we often talked about the leading role the German government had played in enacting and enforcing new laws. Dirk and Tanja wanted us to see firsthand why they and others in Germany cared so much about these issues. As the stewards of vast amounts of personal data, tech companies need to appreciate, perhaps as only people who suffered under the Nazis and Stasi could, the risks of data falling into the wrong hands. "Many of those who came to this prison were arrested for things they did in the privacy of their homes," Dirk said. "It was a system of total surveillance designed to control the people."

The experiences under the Nazis and Stasi, he explained, had made modern-day Germans wary of electronic surveillance. And the Snowden revelations had only fed those suspicions. "If data is collected, it can always be abused," he said. "It's important that, as we operate around the world, we remember that governments can change over time. Look what happened here. Data collected about people—their political, religious, and social views—can fall into the wrong hands and cause all sorts of problems."

Back in Redmond when I talked with employees about privacy, Scheidler's story helped illuminate what was at stake when we handled our customers' data. Privacy wasn't just a regulation that we had to abide by, but a fundamental human right that we had an obligation to protect.

The story also helped people understand that when cloud computing went global, it involved more than laying fiber-optic cables under oceans and building data centers on other continents. It also meant adapting to other countries' cultures while maintaining our commitments to core values by respecting and protecting other people's privacy rights.

A decade ago, some in the tech sector thought they could serve the customers of the world solely from data centers in the United States. But soon real-world experience dispelled this notion. People expected web pages, emails, and documents with photos or graphics to load on their phones and computers instantaneously. Consumer tests showed that a delay of just a half second would get under people's skin.[4] The laws of physics required building data centers in more countries, so this content wouldn't have to travel on cables halfway around the world. This geographic proximity is key to reducing what we call data latency, or the lag in transmission.

Even before our Quincy data center broke ground, we started hunting for a European home for what would become our first data center outside the United States. The early front-runner was the United Kingdom, but soon Ireland entered the race.

Since the 1980s, Ireland had been something of a second home to the American tech sector. Microsoft had been the first technology company to invest there in a big way. Tax incentives and an English-speaking workforce first drew companies to the Emerald Isle. The country then used its membership in the European Union and its welcoming spirit to attract people from across Europe and then from around the world to live and work there, especially in the Dublin area. It fed the Celtic Tiger and sustained a new generation of prosperity for the small country. At Microsoft we took pride in our connections and contributions to this growth.

Back in the 1980s, our European customers installed our software from CD-ROMs, which were manufactured in Ireland. But as software transitioned to the cloud, the Irish realized that the CD business would eventually vanish. They needed to make a new economic bet for the country.

The Irish Department of Enterprise, Trade and Employment did a masterful job of seeing this future and building a foundation that attracted data centers to the country. When they visited me and others in Redmond when the cloud was just a twinkle in our eye, they made the case for putting our first European data center near Dublin. The delegation included a senior official named Ronald Long, whom I'd worked with during my days as a lawyer at Covington & Burling in London. I'd once spent an afternoon hammering out a challenging public policy issue with him in Dublin.

I paused reluctantly in our meeting in Redmond and explained that it just wasn't feasible for us to build our first European data center in Ireland. There was no high-speed fiber-optic cable connecting Ireland to the European mainland, and without that, a data center in Ireland simply didn't make sense.

Ron's answer was simple: "Give us three months."

How could we say no to that?

Three months later the Irish government had negotiated a contract for precisely the type of cable needed. And we were on our way to building a data center south of Dublin. We started with a small building. Then we added more. And more.

In 2010, Microsoft began storing in Ireland our data for customers across Europe. Today we have data centers in several other countries across Europe, but none is as large as our data center campus in Ireland, which matches our biggest facilities in the United States. It fills two square miles. Together with the large data centers run by Amazon, Google, and Facebook, it has helped turn Ireland from a small island into a data superpower.

Ireland today provides one of the world's best locations for data centers. While some may think it's because of tax incentives, other factors are far more important. One is the weather. At a time when data centers are collectively the world's largest consumers of electricity, Ireland's mild climate provides the ideal temperature for computers. The buildings don't need to be cooled, and the heat recirculated from the servers themselves is often all that's needed to warm the buildings in the winter.

But more important than the weather is Ireland's political climate. The nation is both part of the European Union and the beneficiary of a durable local consensus that respects and protects people's human rights. There is a strong but pragmatic data protection agency that understands technology but ensures that tech companies protect the personal information of their users.

As I commented to officials while visiting nations in the Middle East, "Ireland is to data what Switzerland is to money." In other words, it is a place where people should want to store their most precious personal information. It feels like the last place that would produce a modern-day counterpart to the Stasi prison we had walked through in Berlin.

Unfortunately, the global operation of data centers has become far more complicated than simply putting data in a place like Ireland.

One reason is that more countries now want to store their data within their own borders. While this prospect had never excited the tech sector, in some ways it's understandable. In part it's a matter of national prestige. It also guarantees that a government can apply its own laws and ensure that its search warrants can reach all the country's data.

The pressure to put data centers in more countries is giving rise to what rapidly is becoming one of the world's most important human rights issues. With everyone's personal information stored in the cloud, an authoritarian regime bent on broad surveillance can unleash draconian demands to monitor not only what people are communicating, but even what they're reading and watching online. And armed with this knowledge, governments can prosecute, persecute, or even execute those individuals they consider threats.

This is a fundamental fact of life that everyone who works in the tech sector needs to remember every day. We're fortunate to work in one of the most lucrative economic sectors of our lifetime. But the money at stake pales in comparison to the responsibility we have for people's freedom and lives.

For this reason, every decision to put a Microsoft data center in a new country requires a detailed human rights assessment. I review the findings and get personally involved whenever these raise concerns—especially when the final answer needs to be no. There are countries where we have not and will not place data centers because the human rights risks are too high. And even in other nations where the risks are lower, we store business but not consumer data, put in place additional safeguards, and remain vigilant. New demands can suddenly create quiet but dramatic crises. There are days and nights that test the moral courage of those responsible for the cloud.

Even when all this goes well, a second dynamic can undo all the protection that comes from storing data in a place like Ireland. It's when a government in one country seeks to require a tech company to turn over data stored in another. If there is no orderly process that safeguards human rights, then countries all around the world can seek to reach over each other's borders, including into safe havens like Ireland.

In some respects, it's not a new issue. For centuries, governments around the world agreed that a government's power, including its search warrants, stopped at its border. Governments had the authority to arrest people and search homes, offices, and buildings within their own territory, but they couldn't swoop into another country to snatch a person or remove documents. Instead they had to work through the government of that sovereign territory.

There were times when governments ignored this system and instead took matters into their own hands. This disrespect for borders increased international tensions and contributed to the events that eventually led to the War of 1812 between the United Kingdom and the United States. Hostilities between the two countries swelled when the British Royal Navy ruled the seas but was perpetually short of sailors for its naval war against Napoleon. To replenish its depleted crews, the British would send "press gangs" onto foreign ships and into foreign ports to kidnap men and impress them into service. While the theory was that the King's navy was picking up British subjects, the press gangs didn't exactly look for passports. When it was revealed that they were grabbing people indiscriminately and forcing some American citizens into the Royal Navy, the United States demanded action. The young nation barred armed British vessels from calling on American ports entirely. The message was clear: Respect our laws or leave the country.[5]

It would take the War of 1812 before both governments came to their senses and agreed to respect each other's sovereignty. A new field of international treaties emerged that provided for the extradition of criminals and access to information in other countries. Many of these new agreements were called MLATs, or mutual legal assistance treaties.[6] Over the past decade, however, it has become apparent that they're often ill-suited for an era of cloud computing. Law enforcement agencies were understandably frustrated by the slow pace the MLAT process sometimes entailed, but while governments discussed ways to update the agreements and accelerate the process, progress was slow.[7]

As data moved to the cloud, law enforcement agents sought a way to work around the MLAT process. They would try to serve a warrant on a tech company located within their jurisdiction, demanding emails and electronic files that were stored in a data center located in another country. As they saw it, there was no longer a need to rely on an MLAT. They didn't even need to tell the other government what they were doing.

Most governments, however, were understandably less than enthusiastic about having a tech company pull their citizens' data and turn it over to foreigners, bypassing their own legal protections. Back in 1986 when the US Congress had enacted ECPA, it had included a provision that ensured that other countries couldn't do this. They didn't want to see foreigners act like press gangs for digital data. ECPA made it a crime for a US tech company to turn over certain types of digital data such as email, even in response to a legal demand from a foreign government. Similarly, the 1968 Wiretap Act made it a crime to intercept, or wiretap, communications inside the United States for a foreign government. We were required instead to go through an established international process with an MLAT.

Europe's laws were less explicit, but we knew their views mattered as much as those of people in our own country. They didn't like foreign governments reaching into their territory any more than American officials did, especially because the European Union and its members had enacted strong laws to protect their citizens' privacy rights. We knew that like British ships in American ports in the early 1800s, our data centers would be welcome on European soil only if we agreed to respect local laws.

As cloud computing became more ubiquitous and data more accessible, however, the temptation for governments to act unilaterally to seek data in other countries proved irresistible. On an individual, case-by-case basis, this was understandable. A law enforcement investigator needed information and wanted it as quickly as possible. Why take the time to go through a lengthy MLAT process with another government if a tech company with an office down the street could be compelled to act more quickly? If the other government objected, the tech company would end up dealing with the fallout rather than a local prosecutor.

At Microsoft, we soon found ourselves in the middle of these new battles, ducking bullets from both sides. Cases in two countries came to epitomize this challenge.

One country was Brazil. On a January morning in 2015, one of the leaders of our Brazilian subsidiary was in Redmond for a sales meeting when he stepped out into the hallway to answer a call from his wife. She was home in São Paulo and sounded frantic. The Brazilian police had come to arrest him and were demanding that he appear. They'd burst through the building's gates and locked down his apartment. What was his crime? He worked for Microsoft.

The Brazilian police insisted that we turn over personal communications in connection with an ongoing criminal investigation under Brazilian law. But we had no data center in Brazil at the time, and the laws of physics would require this to occur in the United States. We explained that this would constitute committing a crime under US law, and instead we encouraged them to work through the MLAT process in place between the two countries. The Brazilian authorities took a dim view of our suggestion. They had already brought one criminal case against another one of our local executives in São Paulo in a similar situation, and fines against Microsoft were rising on a monthly basis.

We asked Nate Jones to try to negotiate with the Brazilian officials. "We were stuck between a rock and a hard place, and the Brazilian rock didn't want to budge," he later said.

While it was easy for Nate to continue to address the issue from the security of his office in Redmond, our local leaders in Brazil didn't have that luxury. The authorities in São Paulo briefly jailed one of our executives and refused for years to dismiss the criminal charges against him. We readily took on the expenses to defend him in court and said we'd move him and his family out of Brazil if they wanted. We also took on the challenge of appealing more than $20 million in fines against the company.

The second challenge came from the United States. In late 2013, a warrant had arrived demanding email records in connection with a drug trafficking investigation. While that was typical, a review of the account quickly revealed something that was not. These emails appeared to belong to someone who was not a US citizen. And they were stored not on American soil but in Ireland.

We hoped the FBI and DOJ would turn to the Irish government for assistance. After all, the United States and Ireland are close, friendly allies with an updated MLAT in place. We spoke with officials in Dublin and confirmed they were willing to help. But the DOJ officials didn't like the precedent this set for a practice they didn't want to pursue. They said we needed to comply with the warrant.

For us, the precedent was equally important. If the US government could reach into Ireland without regard to Irish law or even having to let the Irish government know, then any other government could do the same thing. And they could try to do this anywhere. We decided to litigate rather than concede.

In December 2013, we took our case to federal court in New York. Our journey to the courthouse building in Foley Square in Lower Manhattan brought me back to my professional roots. I had spent my first year after graduating from Columbia Law School in 1985 working for a district judge on the twenty-second floor of the same narrow building near Wall Street. The clerkship provided an insider's view of the mechanics of the law.

New York felt a long way from the town of Appleton, in northeastern Wisconsin where I had grown up. And while the big city was quite a departure from my Midwestern upbringing, I hadn't appreciated when I arrived for my first morning of work that I also was something of a novelty. I brought not only the eager disposition of a new law school graduate, but a sight that was unusual for the storied courthouse—a new employee carrying a heavy but powerful personal computer.[8]

I had bought my first computer the preceding fall at a time when, for most people, the devices were still uncommon. If truth be told, the soon-to-be-discontinued IBM PCjr wasn't much of a computer. But I had loaded on it a software program that had transformed my final year of law school. It was version 1.0 of Microsoft Word. I loved the software so much that I still have the disks, manual, and plastic case sitting in my home office today. Compared to a pen and paper or the typewriter I had used in college, word processing was like magic. Not only could I write faster, I could write better. So I persuaded my wife, Kathy, a new lawyer herself, that before starting my first job, I should spend ten percent of my annual $27,000 salary to buy a better PC and install it in my office at work. Thank goodness she was so supportive.

The judge for whom I worked was seventy-two years old at the time, and the office with my desk was filled with shelves of well-organized boxes containing his meticulous handwritten notes from more than two decades of trials and cases. There was an elaborate—and time-tested—filing system with typed cards for each of the points that needed to be assembled for jury instructions. My arrival with a personal computer raised some eyebrows. That's when I first realized the importance of using my computer to do what I needed to do better—writing memos and drafting legal decisions—without upsetting old practices that still worked well. It's a valuable lesson that I take with me to this day: Use technology to improve what can be improved while respecting what works well already.

Fast-forward to 2014, and once again we were injecting new computing technology into the same courthouse. We knew we likely faced a long battle, a view that was quickly confirmed when a local magistrate judge ruled against us, setting the stage for a lengthy climb up a tall appellate ladder.

The public response to our case was swift, especially across Europe. A month after our defeat, I traveled for a series of meetings that started in Berlin with government officials, members of Parliament, customers, and reporters. While I knew our Irish warrant case would be of interest, I hadn't expected the intensity and consuming focus on the case. In truth, as I began the first morning at eight o'clock talking with a reporter, I struggled before my second cup of coffee to recall the name of the magistrate judge who had issued the initial decision against us. Our litigation team had already recovered from the blow, picked itself up, and was warming up for round two before the district judge. We'd moved on, but I quickly learned that the Germans had not.

By the end of my two days in Berlin, the minutiae of the decision and the name of the magistrate judge who had written it were seared in my memory. Everywhere I went, people almost immediately started talking to me about Judge Francis. Almost no one outside a small legal circle in New York had heard of him, but in Berlin in 2014, James C. Francis IV, the magistrate judge who had ruled against us, had become a household name.

The questions seemed endless. "What did he mean by . . . ? Why did he say . . . ? What happens next?" The Germans brought copies of Judge Francis's decision that they'd carefully annotated. A few people read passages aloud to me. Many had studied every page.

By the time I sat down the first afternoon with the chief information officer for one of Germany's largest states, I was weary. The CIO laid Judge Francis's decision on the mahogany table between us. He ground his index finger squarely into the legal decision and declared, "There is absolutely no way that my state will ever put any of our data in an American company's data center unless you get this reversed."

The issue followed us on our international travels that entire year. In Tokyo, I had not expected the same reaction I'd experienced in Berlin. But at a reception I was besieged by a crowd of enterprise customers determined to tell me in person how important the outcome of our Irish data center case was to their business. "Microsoft must win this case," they said again and again. They too would watch closely as our case worked its way through the courts. At public appearances around the world, I repeatedly vowed that we would stick with the case and attempt to take it all the way to the Supreme Court if needed.

As the case slowly ground forward, we recognized that even if we won, the lawsuit had its limitations. It could call the question on the reach of search warrants under existing law, but it could never put in place a new law or the new generation of international treaties needed to move past outdated MLAT agreements.

We started to draft new proposals and walk the halls of government offices around the world in search of allies who might spearhead the broader initiatives needed. Legislation was introduced in Congress,[9] but we also needed to couple this with new international agreements.

In March 2015, we caught a break. A meeting I attended at the White House created an opportunity to review ongoing privacy and surveillance issues. As I described the criminal case against our Brazilian executive and the fines against Microsoft, President Obama interrupted and observed, "This sounds like a mess." The group discussed and the president endorsed the opportunity to develop a new approach to international agreements, preferably with one or two key allied governments like the United Kingdom or Germany.

Eleven months later, in February 2016 and with little fanfare, the UK and US proposed the draft for a more modern bilateral data-sharing agreement. One of our building blocks was now emerging. But the agreement couldn't go into effect without a new statute passed by Congress, and despite broadening endorsements across Capitol Hill, the DOJ continued to balk at any legislation that would change the way it used search warrants to obtain data around the world. We faced a legislative stalemate, and without a broader compromise it was difficult to be optimistic about our prospects.

As it turned out, the Supreme Court itself broke the stalemate, and in an unlikely way.

It would take until late February 2018, but on an unseasonably warm morning we walked down First Street in Washington, DC, toward the towering pearly facade of the United States Supreme Court.[10] We paused to take in the magnificent sight where the global implications of cloud computing would be presented to the Court's nine justices.

The Supreme Court's majestic four-story building sits directly across from the US Capitol—the physical intersection of the American judicial and legislative branches. Look one direction and the Capitol's gleaming dome fills the sky. Turn around and you'll gaze up the stretch of deep marble steps, past soaring columns, toward a pair of tall carved doors that mark the court's entrance.

When we arrived on February 27, a long trail of people snaked down the iconic staircase and around the block, a line of hopeful observers waiting to watch us face off against our own government. This would be the final judicial showdown in a battle that had started four years earlier when we refused to move email across the Atlantic from Ireland.

It was the fourth time Microsoft had argued a case before the Supreme Court. I've always found it a striking experience. We bring the issues created by the world's most modern technology into a courtroom that looks the way it did almost a century ago. No phones and no laptops are permitted. Each time, after I leave my devices behind, I take my seat in the massive red chamber that resembles a curtained stage. I then gaze up at the courtroom's lone piece of technology: a clock.

I've come to appreciate the Supreme Court's ability to consider technology's implications in a setting where no modern technology is in view. Our first case before the court, back in 2007, involved patent issues that arose, coincidentally, from our CD manufacturing in Ireland.[11] A week after the argument, I encountered one of the court's senior administrators, who said, "You looked a little dismayed when some of the justices were speaking."

I realized that I clearly had not done a good job of keeping a poker face. I still remember the occasion. At the time, one of the justices was discussing with opposing counsel the implications of Microsoft "sending photons" from New York to computers in Europe.[12]

"What has this case got to do with photons?" I wondered. "And why are we talking about New York?"

But I had learned a valuable lesson that went beyond the need for me to keep a straight face during the hearing. The justices didn't always understand every detail of the latest technology, but they had younger clerks who did. And the justices complemented that factual understanding with wisdom and judgment that often went even beyond the law itself. Despite the public rancor over nominations and certain controversial cases, the Supreme Court remains one of the world's truly great institutions. Most days nine justices try to reason through challenging problems together. I've been in courtrooms around the world and have developed a confidence in what the US Supreme Court can accomplish.

On this morning, after an hour of oral argument, the court's nine justices left both sides with less confidence than either of us would have liked. While there was ample room to speculate on who might triumph, it was impossible to make a prediction with great confidence. Whether by accident or by design, the justices created the perfect atmosphere to encourage both sides to reach a settlement.

But there remained a huge hurdle. Only if a new law was passed could both sides agree that a Supreme Court ruling was no longer necessary. In other words, a settlement required new legislation that could only come from the other side of First Street, the Capitol.

At one level, asking for an act of Congress felt like asking for an act of God. Congress was divided on almost everything, and it wasn't in the habit of passing much legislation. But we saw a small window of opportunity. I talked through our options with Fred Humphries, a long-standing Washington hand who leads our government affairs team. Together with the White House, we decided to try.

It would not have been conceivable without bipartisan efforts in both the Senate and House of Representatives that had begun shortly after we had filed the lawsuit four years before. But after two legislative hearings and a slew of iterations, we sat down for a final round of discussions with the DOJ that was brokered by Senator Lindsey Graham, using his position as chairman of the Senate Judiciary Subcommittee on Crime and Terrorism.

Graham acted with determination to encourage people to come together. He had held a well-attended hearing at which I had testified almost a year earlier, in May 2017. The British government sent its deputy national security adviser, Paddy McGuinness, to testify as well, given the implications for its international agreement with the United States. He combined an amiable Scottish disposition with a pragmatic but hardheaded understanding of what it took to fight terrorism in the UK. White House Homeland Security adviser Tom Bossert talked regularly with McGuinness and pushed everyone to find common ground in Congress.

Following the Supreme Court's argument, there soon emerged a new text that both sides agreed to support. It was given a new name, the Clarifying Lawful Overseas Use of Data Act, or CLOUD Act.

The legislation had provisions that we cared about. It balanced the international reach for search warrants that the DOJ wanted with a recognition that tech companies could go to court to challenge warrants when there was a conflict of laws. This meant that if Ireland, Germany, or the entire European Union wanted to block unilateral foreign search warrants through their local laws and instead compel a more transparent or collaborative approach, they could do so and we could rely on this in a US courtroom.

Even more important, the CLOUD Act created the new authority for modern international agreements that could replace these unilateral efforts. These agreements can enable law enforcement agencies to access data in another country with faster and more modern procedures, but with rules to protect privacy and other human rights. Like all legislation, especially those that involve compromise, it wasn't perfect. But it had most of what we had spent more than four years trying to advance.

But finding a vehicle to get the CLOUD Act passed remained a big problem. It was unlikely that both the Senate and the House would have time on their legislative calendars to take up the issue by itself, especially in the short time before the Supreme Court ruled. We'd need to attach it to another piece of legislation.

We recognized that the only real prospect for passage would be to attach the proposal to a budget bill. This would be a difficult stretch for two reasons. First, Congress was having a hard time passing budget bills. And second, for that very reason congressional leaders had become reluctant to attach non-budget proposals to budget bills.

It was apparent, however, that with Senator Graham's support Senate Republicans might support the idea. But it would go nowhere if Senate Democrats balked. We knew immediately that there was one person who could make a difference. In many ways, we regarded him not only as a legislative leader but also as a veritable force of nature—Senate Minority Leader Chuck Schumer. While he was only remotely familiar with the issue, he studied it quickly and took the cause on board.

With Bossert, Graham, and Schumer all pushing, there ensued a feverish effort to bring along leaders in the House of Representatives. Soon both Speaker Paul Ryan and Minority Leader Nancy Pelosi were steeped in discussions about whether to include the CLOUD Act in the budget bill. Negotiations led to another round of amendments. Every couple of days it seemed as if the effort was almost dead, but we talked repeatedly with Bossert and we all resolved not to give up. Remarkably, after many rounds of calls and conversations, it stayed alive. And on March 23, 2018, President Donald Trump signed an omnibus budget bill that included it. The CLOUD Act was now law,[13] and the Supreme Court case would shortly be settled.

It had been more than four years since we had first gone to the federal courthouse in New York. But it had been less than a month since we had left the steps of the Supreme Court. The final stages had proceeded so quickly that it took by surprise even those of us who had been involved in every detail.

While we were pleased with the result, it also made for some mixed feelings. We believed the CLOUD Act made for a strong law. But as with all legislation and all court settlements, it also had elements of compromise. One thing we had learned long ago was that it was more fun to fight a battle but typically more rewarding to strike a deal. It usually was the only way to make progress. And deals required some give-and-take.

They also required that we do a good job of explaining the result, especially when it was complicated. That was one reason we typically planned for a variety of outcomes and had communications material ready to go. But the CLOUD Act had moved so quickly and required so much time talking with people in Washington that we were less prepared for this aspect than we should have been.

Inquiries started to pour in from customers, privacy groups, and government officials around the world about what the CLOUD Act said and how it would really work. Customers had questions, and privacy groups expressed concerns. We scrambled, and soon we were providing briefings around the world and publishing material to fill the gap.[14] It was an exercise that involved Microsoft sales reps in almost every country—a fact brought home as I was stopped on the street a month later in France by a local Microsoft employee who recognized me as I walked by the restaurant where he was eating dinner. His dinner grew cold as he chased me, recovered his breath, and peppered me with questions about the new law.

The outcome reflected both how far we had come and how far the world still needs to go. There is now a framework for a different future based on new agreements between nations. As US Assistant Attorney General Richard Downing said on the CLOUD Act's first anniversary, the law "offers not simply a solution to the challenge of this moment, but also an aspirational kind of solution." As he explained, it's "a solution aimed at fostering a community of like-minded, rights-respecting countries that abide by the rule of law—where countries can minimize their conflicts of law and advance their mutual interests based on shared values and mutual respect."[15]

But the CLOUD Act is like a foundation on which new houses must be built. We live in a world where law enforcement must move quickly, privacy and other human rights need protection, and countries' borders deserve respect.

New international agreements can accomplish all of these if they're put together thoughtfully and pursued with persistence.

In other words, years of work still lie ahead.

Chapter 4 » **CYBERSECURITY: The Wake-up Call for the World**

On May 12, 2017, Patrick Ward was wheeled into a surgical prep room at St. Bartholomew's Hospital in central London. The sprawling medical complex, known simply as Bart's to locals, sits a few blocks from St. Paul's Cathedral, where it was founded in the year 1123, when Henry I ruled the kingdom. The hospital operated continuously throughout Hitler's aerial bombing campaign, proudly weathering the Second World War as bombs rained around and sometimes even on the elegant landmark.[1] But in Bart's nine hundred years of existence, no bomb created greater havoc than the one dropped that Friday morning.

Ward had traveled three hours from his small village in Dorset, near the southern English town of Poole. His family had farmed the maritime land since the late 1800s, a sweeping landscape plucked straight from the pages of a storybook. And Ward had a job that suited his idyllic home. He was the longtime sales director of Purbeck Ice Cream, a gourmet ice cream maker. He loved the job. "I get paid to talk and eat ice cream," he told us. "And I can do both of those fairly well."

He'd waited two years for a surgical slot to open up at Bart's to repair a serious heart condition called cardiomyopathy, a genetic condition that had thickened his heart wall and rendered the sturdy middle-aged Englishman, who'd hiked and played soccer, unable to perform most daily tasks. That morning, Ward's chest was shaved and his body beset by a battery of tests. He lay resting on a gurney, awaiting his long-sought procedure, when the surgeon dropped in. "It'll be just a few more minutes. I'll see you soon in the back." But Ward was never wheeled into the operating room. He just continued waiting.

More than an hour later, the doctor reappeared. "We've been hacked. The whole system is down across the hospital. I can't do the operation." The hospital that had remained open throughout World War II was suddenly frozen, a target of a sweeping cyberattack. All its computer systems had crashed. Ambulances were diverted, appointments were canceled, and surgical services closed for the day. The attack had paralyzed a third of the country's National Health Service, which provides most of the health care in England.[2]

Later that morning in Redmond, Microsoft's senior leadership team was in the middle of its regular Friday meeting. These weekly gatherings with Satya Nadella and his fourteen direct reports have a routine. They begin at eight a.m. in the company's boardroom, on the floor where several of us have our offices. We cycle through various discussions on product and business initiatives before adjourning midafternoon. But May 12, 2017, was not a typical meeting.

Before we'd finished the second topic, Satya interrupted the discussion. "I'm getting copied on a bunch of emails about a widespread cyberattack on our customers. What's going on?"

We quickly learned that Microsoft's security engineers had been scrambling in response to customer calls and were trying to find the cause and assess the impact of the rapidly spreading attack. By lunchtime, it was clear that this was no ordinary hack. Engineers in the Microsoft Threat Intelligence Center, which we call MSTIC (pronounced "mystic"), quickly matched the malware to code that a group called Zinc had experimented with two months earlier. MSTIC gives each nation-state hacking group a code name based on an element from the periodic table. In this instance, the FBI had connected Zinc to the North Korean government. It was the same group that had overwhelmed Sony Pictures' computer network a year and a half earlier.[3]

Their latest attack was unusually sophisticated from a technical perspective, with new malware code added to the original Zinc software that allowed the infection to worm its way automatically from computer to computer. Once replicated, the code encrypted and locked a computer's hard disk, then displayed a ransomware message demanding three hundred dollars for an electronic key to recover the data. Without the key, the user's data would remain frozen and inaccessible—forever.

The cyberattack started in the United Kingdom and Spain. Within hours it spread around the world, ultimately impacting three hundred thousand computers in more than 150 countries.[4] Before it ran its course, the world would remember it by the name WannaCry, a malicious string of software code that not only made IT administrators want to cry but served as a disturbing wake-up call for the world.

The *New York Times* soon reported that the most sophisticated piece of the WannaCry code was developed by the US National Security Agency to exploit a vulnerability in Windows.[5] The NSA had likely created the code to infiltrate its adversaries' computers. The software was apparently stolen and offered on the black market through the Shadow Brokers, an anonymous group that posts toxic code online to wreak havoc. The Shadow Brokers had made the NSA's sophisticated weapon available to anyone who knew where to find it. While this group has not been linked definitively to a specific individual or organization, experts in the threat intelligence community suspect that it is a front for a nation-state bent on disruption.[6] This time, Zinc had added a potent ransomware payload to the NSA code, creating a virulent cyberweapon that was ripping through the internet.

As one of our security leaders put it, "The NSA developed a rocket and the North Koreans turned it into a missile, the difference being the thing at the tip." Essentially the United States had developed a sophisticated cyberweapon, lost control of it, and North Korea had used it to launch an attack against the entire world.

A few months earlier, this plot had seemed implausible. Now it was the news of the day. But we didn't have time to dwell on the irony of the situation. We had to scramble to help customers identify what systems were affected, stop the malware from spreading, and resuscitate computers that were disabled. By midday, our security team concluded that newer Windows machines were protected against the attack by a patch we had released two months earlier, but older machines running Windows XP were not.

This was not a small problem. There were still more than a hundred million computers in the world running Windows XP. For years we had tried to persuade customers to upgrade their machines and install a newer version of Windows. As we pointed out, Windows XP had been released in 2001, six years before the first Apple iPhone and six months before the first iPod. While we could release patches for specific vulnerabilities, there was no way technology this old could keep pace with current security threats. Expecting sixteen-year-old software to defend against today's military-grade attacks was like digging trenches to defend against missiles.

Despite our urgings, discounts, and free upgrades, some customers stuck with the old operating system. As we sought to move the installed base forward, we eventually decided that we would continue to create security patches for the older systems, but unlike with newer versions, we required that customers purchase them as part of a subscription service. Our goal was to create a financial incentive to move to a more secure version of Windows.

While the approach made sense in most circumstances, the May 12 attack was different. The "wormable" nature of WannaCry enabled the malware to move at unusual speed. We had to stem the damage. This led to a vigorous debate within Microsoft. Should we make the Windows XP patch for this attack available to everyone worldwide, outside the security subscription, including for computers that ran pirated copies of our software? Satya cut off the debate, deciding that we would release the patch to everyone free of charge. When some inside Microsoft objected, saying the move would undercut the effort to move people off XP, Satya quelled the dissent with an email saying, "Now is not the time to debate this. This is too widespread."

While we made technical progress to contain and suppress the WannaCry infection, the political ramifications were heating up. By dinnertime in Seattle that Friday, it was Saturday morning in Beijing. Officials in the Chinese government reached out to our team in Beijing and emailed Terry Myerson, who led the Windows division, asking about the status of patches for Windows XP.

We weren't surprised by the inquiry, given that China had more Windows XP machines than any other nation. China had been mostly spared from the initial attack because most office computers had been turned off for the weekend when the malware was unleashed on Friday evening local time. But their dated Windows XP machines remained vulnerable.

But XP patches weren't the only thing on China's mind. The official who emailed Terry asked about a point made that same day in the *New York Times*. The story claimed that the US government had been searching for and stockpiling software vulnerabilities, keeping them secret rather than notifying tech companies so they could be patched.[7] He wanted our reaction. We said it was a question that they should discuss with the US government, not us. But not surprisingly, it was not a practice we or other tech companies were enthusiastic about. To the contrary, we had long urged governments to disclose vulnerabilities they had identified so they could be fixed for the common good.

We knew that this was just the first of many questions to come from people around the world. By Saturday morning, we realized that we needed to do more than support our affected customers. We needed to address the emerging geopolitical issues more publicly. Satya and I spent some time on the phone that morning discussing our next step. We decided to go public to address the coming wave of questions about WannaCry.

We stepped back from the specifics of the attack and addressed the broader cybersecurity landscape. We said explicitly that Microsoft and other companies in the tech sector had the first responsibility to protect our customers against cyberattacks. This was a given. But we thought it was important to underscore how cybersecurity had become a shared responsibility with customers. We needed to make it easier for customers to update and upgrade their computers, but one of the episode's lessons was that our advances would do little good if they weren't used.

We also made a third point, one that we thought the WannaCry attack had made clear. As more governments developed advanced offensive capabilities, they needed to control their cyberweapons. As we said, "An equivalent scenario with conventional weapons would be the US military having some of its Tomahawk missiles stolen."[8] The fact that cyberweapons could be stored—or stolen—on a thumb drive made their safeguarding both more difficult and more important.

Some officials at the White House and NSA were less than enthused by our reference to a Tomahawk missile. Some of their counterparts in the British government agreed with them, arguing, "It's more accurate to compare WannaCry to a rifle than a Tomahawk missile." But has a rifle attack ever hit targets in 150 countries simultaneously? All of this was beside the point. If anything, it reflected the degree to which cybersecurity officials were unaccustomed to talking directly about these issues in the press or defending their practices to the public.

What surprised us most was the absence of any broad discussion about why North Korea had launched the attack in the first place. To this day we don't have a definitive answer, but one theory is especially intriguing.

Just a month before the attack, the North Koreans experienced an embarrassing failure of a high-profile missile launch. As David Sanger and two other reporters at the *New York Times* wrote at the time, the US government had been seeking to slow the missile program, "including through electronic-warfare techniques."[9]

As they noted in the *Times*, it was impossible to know what caused any specific missile failure, but Defense Secretary James Mattis had been cryptic in commenting on this one, saying only that "The president and his military team are aware of North Korea's most recent unsuccessful missile launch. The president has no further comment." This from a president who is seldom known for having no further comment.

What if the North Koreans had responded to a cyberattack on their missile by retaliating with a cyberattack of their own? WannaCry was indiscriminate in its effect, but what if that was the point? What if it was their way of saying, "You can hit me in one particular place, but I can hit back everywhere"?

Several aspects of WannaCry are consistent with this theory.

First, it was launched against targets in Europe just about the time that everyone in east Asia was turning off their computers and going home for the weekend. If the North Koreans wanted to maximize the impact on Western Europe and North America while reducing the impact in China, they chose the ideal moment. The infection spread as the sun moved west, as employees in businesses and governments on other continents continued their workday. But the Chinese had the weekend to respond before returning to work on Monday.

In addition, the North Koreans had added what security experts call "kill switches," which made it possible to stop the malware from spreading further. One kill switch directed the malware to look for a specific web address that did not yet exist. As long as it wasn't there, WannaCry would continue to spread. But once someone registered and activated the web address, which was a simple technical step, the code would stop replicating.

Late on May 12 a security researcher in the United Kingdom analyzed the code and found this kill switch. For the modest price of $10.69, he registered and activated the URL, stopping WannaCry from spreading further.[10] Some speculated that this reflected a lack of sophistication by WannaCry's creators. But what if the opposite was true? What if WannaCry's designers wanted to ensure that they could turn off the malware before Monday morning, so they could avoid causing too much disruption in China or North Korea itself?

Finally, there was something fishy about the ransomware message and approach used by WannaCry. As our security experts noted, North Korea had used ransomware before, but their tradecraft had been different. They had selected high-value targets such as banks and demanded large sums of money in a discreet way. Indiscriminate demands to pay three hundred dollars to unlock a machine represented a departure, to say the least. What if the whole ransomware approach was just a cover to throw the press and public off the real message, which was intended to be more discreetly understood by US and allied officials?

If North Korea was responding with its own cyberattack to a US cyberattack, then the whole episode was even more significant than people have appreciated. It was the closest thing the world has experienced to a global "hot" cyberwar. It would mean that this was an attack in which the impact on civilians represented more than collateral damage. It was the intended effect.

Regardless of the answer, the question reflects a serious issue. Cyberweapons have advanced enormously over the past decade, redefining what is possible in modern warfare. But they are used in ways that obscure what is truly happening. The public doesn't yet fully appreciate either the risks or the urgent public policy issues that need to be addressed. And until these issues are brought out of the shadows, the danger will continue to grow.

If anyone doubted the threat of cyberwarfare, the cyber bomb that hit just six weeks later should have made them believers.

On June 27, 2017, a cyberattack pummeled Ukraine, using the same software code stolen from the NSA, disabling an estimated 10 percent of all computers in that country.[11] The attack was later attributed by the United States, the United Kingdom, and five other governments to Russia.[12] Security experts dubbed it NotPetya because it shared code with a known ransomware named after the armed satellite *Petya*, which was part of the Soviet Union's fictional GoldenEye weapon in the 1995 James Bond movie of the same name.[13] That weapon could knock out electronic communications across a thirty-mile radius.

In the nonfiction world of 2017, the NotPetya attack had a far broader reach. It rippled across Ukraine, crippling businesses, transit systems, and banks, then continued spreading outside the country's border, infiltrating multinationals including FedEx, Merck, and Maersk. The Danish shipping giant saw its entire worldwide computer network grind to a halt.[14]

When Microsoft security engineers arrived at Maersk's offices in London to help revive their computers, what they encountered was eerie by twenty-first-century standards. Mark Empson, a tall, fast-talking, fast-moving Microsoft field engineer, was one of the first on the scene. "You always hear noise and ambient sound effects from computers, printers, and scanners," he said. "Here there was nothing at all. It was absolutely silent."

As Empson walked Maersk's hallways, he said it felt as if the office had died. "You go through the standard troubleshooting logic of, 'Okay, well, what's the situation? What servers are out? What have we got?' The answer was that everything's out." He continued to quiz people. "'Okay, telephones?' 'Out.' 'What about the internet?' 'No, that's out as well.'"

It was a stark reminder of the degree to which our economy and lives rely on information technology. In a world where everything is connected, anything can be disrupted. That's part of what makes it so serious to contemplate a cyberattack targeting today's electronic grid.

If a city loses its electricity, telephones, gas lines, water system, and internet, it can be thrown back into something that can feel like the Stone Age. If it's winter, people may freeze. If it's summer, people may overheat. Those who rely on medical devices to survive could lose their lives. And in a future with autonomous vehicles, imagine a cyberattack that penetrates automobile control systems as cars barrel down the highway.

All these are sobering reminders of the new world we live in. Following NotPetya, Maersk took the unusual step of reassuring the public that its ships remained under the control of their captains. The need for such reassurance illustrated the extent of the world's reliance on computers and the potential for cyberattack-based disruption.

This ubiquity of software in the infrastructure of our societies also explains part of the reason more governments are investing in offensive cyberweapons capabilities. Compared to early teenage hackers and their successors who operate in international criminal enterprises, governments operate on a completely different scale and level of sophistication. The United States was an early investor and remains a leader in the space. But others are quick studies, including Russia, China, North Korea, and Iran, who have all joined the cyberweapons race.

The WannaCry and NotPetya attacks represented a massive escalation of the world's growing cyberweapon capabilities. Yet just a few months later, it became apparent that the world's governments were not necessarily heeding this wake-up call.

In conversations with diplomats around the world, we heard the same skepticism: "No one has been killed. These aren't even attacks on people. They're just machines attacking machines."

As we also found, perhaps more than any prior advance in weapons technology, views about cybersecurity fall along generational lines. Younger generations are digital natives. Their entire lives seem to be powered by technology, and an attack on their device is an attack on their home. It's personal. But older generations don't always see the impact of a cyberattack the same way.

This leads to an even more sobering question. Can we wake up the world before a digital 9/11? Or will governments continue to hit the snooze button?

After NotPetya, we wanted to show the world what had happened on the ground in Ukraine, a country that had suffered multiple cyberattacks. Although the country had been devasted by NotPetya, media coverage outside Ukraine was sparse. We decided to send a team of Microsoft employees to interview people in Kiev and find out what really happened.[15] They captured firsthand accounts of people who had lost their businesses, their customers, and their jobs. They talked to Ukrainians who weren't able to buy food because credit cards and ATMs stopped working. They spoke to mothers who couldn't find their children when a communications network failed. It obviously wasn't a 9/11, but it showed where the world was heading.

The Ukrainians were open about their experiences, but often the victims of cyberattacks remain silent because they're embarrassed about their network security. This is a recipe that will perpetuate rather than solve the problem. Microsoft has faced this same dilemma. In 2017, our lawyers raised the prospect of prosecuting two criminals in the United Kingdom who had successfully hacked into part of our Xbox network. Although this issue raised a few awkward questions, I gave the green light to go public. We could never hope to offer public leadership if we didn't exercise more courage ourselves.

But we not only needed to say more, we needed to do more.

Diplomats across Europe agreed. "We know there's more to be done, but we don't yet know what we should do," one European ambassador said to me at the United Nations in Geneva. "And even if we did, it's not easy to get governments to agree on anything at the moment. This is an issue where tech companies will need to lead. That's the best way to get governments to follow."

Soon this opportunity emerged. A group of security engineers concluded that if a handful of companies acted together and simultaneously, we could dismantle an important part of the malware capability of the North Korean group—or Zinc—that was responsible for WannaCry. We could distribute patches for the vulnerabilities Zinc was seeking to exploit, clean up impacted PCs, and turn off the accounts across our collective services that the attackers were using.

The impact would not be permanent, but it would inflict a blow to the group's capabilities.

At Microsoft, Facebook, and elsewhere, we talked at length about whether and how to move forward. This move would make us a bigger target. I talked it through with Satya and in November we shared with the Microsoft board of directors our plan to go forward. We concluded that we were on strong ground, both legally and otherwise, and if we acted with other companies, it was a step worth taking.

We also concluded that we needed to inform the FBI, NSA, and other officials in the United States and other countries. We would not ask their permission, but simply inform them of our plans. We wanted to make sure that there wasn't an intelligence operation underway to address the North Korean threat based on the same user accounts we were disabling.

A few days later, I was in Washington, DC, and spent part of the day at the White House. I met in the basement offices of the West Wing with Tom Bossert, the president's Homeland Security adviser, and Rob Joyce, the White House cybersecurity coordinator. I told them of our plans, which by then had been set for the following week.

They shared that, with President Trump's strong support, they were close to formally attributing the WannaCry attack to North Korea. It was a key step in starting to hold governments accountable publicly for cyberattacks. Bossert had concluded that it was important for the US government to formally voice its opposition to cyberattacks that it regarded as "disproportionate and indiscriminate." And in this case, the White House was actively working with other countries so that, for the first time, they could all stand together to point the finger publicly at North Korea.

Bossert first asked us to postpone. "We won't be ready to make our announcement until a week later, and it would be better if we all went at the same time." I said we couldn't delay our operation because it had to be timed with the release of certain patches that the public expected on December 12. It was well known that we released patches on the second Tuesday of every month, on what we call Patch Tuesday.

I offered an alternative. "Let's explore postponing our communications about our operation, and maybe we can pursue that together."

The conversation surfaced an important but ironic aspect of the governmental response to WannaCry. As Bossert explained to me and would later say at a press briefing, the US government had only limited responses available in terms of what it could do in response to the incident, given all the sanctions that were already in place. "President Trump has used just about every lever you can use, short of starving the people of North Korea to death, to change their behavior," he said publicly.[16]

While the Administration later would conclude that it could expand its potential responses to cyberattacks, the tech sector was in a position to take some steps that the government could not. Tech companies could readily dismantle certain key parts of North Korea's malware capability. So, it would make a stronger statement to nation-state actors if we could couple the two announcements.

We set out on two coordinated paths, one to execute the operation and the other to announce it publicly. The security teams at Microsoft, Facebook, and another tech company that didn't want to be named publicly worked side by side to disrupt Zinc's cyber capabilities on the morning of December 12, Patch Tuesday. The operation proceeded without a hitch.

But the effort to announce the operation was more complicated. Security professionals in almost any field are typically reluctant to talk publicly about what they do. In part, this is because their culture is about shielding rather than sharing information. And there's always some risk that action will encourage retaliatory attacks. But we had to overcome this reluctance if we were to take effective action against nation-state cyberattacks.

We also faced the added complexity of the tech sector's complicated relationship with the Trump White House. We'd spent recent months in renewed immigration battles, among other issues. This made some people reluctant to acknowledge publicly that they were doing anything with the administration. But I felt that we needed to partner where we could even if we needed to stand apart where we should. And cybersecurity was a common cause in which we not only could partner but needed to work together if we were to make real progress.

The White House said it would make its announcement on December 19. We quickly told Facebook and the other company that we would be willing to make our action against Zinc public if we all wanted to move forward together. But on the morning before, we were standing alone, still waiting for the other two to decide. If necessary, I determined that we would go alone. I believed that the only way we'd effectively deter countries from engaging in these attacks was if we showed that there was a growing ability to respond. Somebody had to be the first to step forward. It might as well be us.

That evening, the good news arrived that Facebook would go public with us and talk about our collective action. And even better news came the next morning when Bossert spoke at a White House press briefing. He explained that the United States was joined by five other nations—Australia, Canada, Japan, New Zealand, and the United Kingdom—in its public attribution. It was the first time countries had come together multilaterally to hold another nation publicly accountable for a cyberattack. He then announced that Microsoft and

Facebook had taken concrete steps the previous week to dismantle part of the group's cyberattack capability.

Acting together, governments and tech companies had accomplished more than any of us could have separately. It was hardly a panacea for the world's cybersecurity threats. It was barely a victory. But it was a new beginning.

Chapter 5 >> **PROTECTING DEMOCRACY:**
"A Republic, If You Can Keep It"

In 1787, as the American constitutional convention reached its conclusion in Philadelphia, Benjamin Franklin was asked as he departed Independence Hall what type of government the delegates had created. He famously replied, "A republic, if you can keep it."[1] That statement would echo across the country and through the ages. It underscored that a democratic republic was not just a new form of government, but one that would require vigilance and at times action to protect and maintain it.

For much of American history, that has meant action by citizens in the form of voting, public service, and, at times, even laying down their lives. At other pivotal moments, it has required the country's businesses to mobilize, like the American industrial sector did to help win World War II. History has taught us that vigilance must be constant because the need to act will arise unexpectedly.

The need for this vigilance arose suddenly late one Sunday night in July 2016. I had spent much of the preceding two weeks attending the Republican and Democratic national conventions in Cleveland and Philadelphia. My weekend had been consumed by catch-up work and I was about to call it a night when an email arrived marked "urgent." As I clicked it open, I could never have imagined that it was the opening salvo in an expanding campaign that would test the tech sector and challenge the industry to step up and defend democracy.

The email was from Tom Burt, a deputy general counsel at Microsoft. The subject line read "Urgent DCU Issue." DCU is Microsoft's Digital Crimes Unit, one of the teams Tom managed. We created it fifteen years ago, and somewhat to my surprise it has remained unique in the tech sector. It comprises more than a hundred people around the world and includes former prosecutors and government investigators, as well as top-tier forensic, data, and business analysts. The DCU was born out of our anti-counterfeiting work in the 1990s but evolved into a digital SWAT team to work with law enforcement when we saw new forms of criminal activity begin to proliferate on the internet.[2]

Ten days earlier, on the Friday before the 2016 Democratic National Convention, WikiLeaks had released emails stolen by Russian hackers from the Democratic National Committee, or DNC. It became a major news story throughout the week of the convention. As the week progressed, our threat intelligence center, MSTIC, identified a new and separate hacking attempt by Strontium—our name for a Russian hacking group also known as Fancy Bear and APT28. Tom's team wanted to launch a legal assault that coming Tuesday to disrupt Strontium.

The FBI and intelligence community had connected Strontium with the GRU, Russia's military intelligence agency. Tom reported that Strontium was spoofing Microsoft services to target a variety of political officials and candidates, including accounts belonging to the DNC and Hillary Clinton's presidential campaign, throwing us into the middle of the story.

MSTIC had monitored Strontium since 2014 as it engaged in so-called spear-phishing attacks, sending carefully crafted emails as a ploy to trick targets into clicking on links from what appeared to be trusted websites, some of which contained the name Microsoft. Strontium would then use a variety of sophisticated tools to engage in key logging, email address and file harvesting, and information gathering from other computers. The group even used a tool to infect connected USB storage devices to try to retrieve data from other air-gapped computers that were not on the network.

Strontium was not only more sophisticated but also more persistent than criminal hacking enterprises, sending a selected target numerous phishing emails over an extended period. Successfully scamming a high-value target was clearly worth the investment.

Even though this type of tactic is known to many computer users, it's hard to combat. As one person tweeted from network security company RSA's annual conference in San Francisco, "Every organization has at least one employee who will click on anything." The technique takes advantage of human curiosity, as well as people's carelessness. As we analyzed hackers' activities, we found that the first thing they often did when they successfully penetrated an email account was search for the keyword *password*. As people accumulated more passwords for more services, they often sent emails to themselves with the word *password* in them, which made for easy pickings for hackers.

In July 2016, MSTIC detected attempts by Strontium to register new internet domains to steal user data. Strontium had started using the Microsoft name in these domains—for example, Microsoftdccenter .com—to make the link look like a legitimate Microsoft support service. The DCU had spent the weekend developing a legal strategy to address the problem, and as Tom reported on Sunday, they were ready to pursue their plan to shut down these sites.

This plan was built on both a legal and a technical innovation the DCU had previously pioneered. We would go to court and assert that Strontium was infringing on the Microsoft trademark, and on that basis ask that control of the new internet domain be transferred to the DCU. In a sense, that part was innovative, but it was also relatively obvious. Trademark law has been around for decades, and these days it prohibits someone from incorporating without permission a registered trademark like "Microsoft" into a website's name.

From a technical perspective, we would then create within the DCU's forensic lab a secure "sinkhole" separate from the rest of our network. This sinkhole would intercept all the communications sent back from infected computers intended for Strontium's command-and-control server. The goal was to take control of Strontium's network, identify which customers had been infected, and then work with each user to clean up their infected devices.

I loved the idea. It was a classic example of why we'd set up the DCU in the first place—to get our lawyers and engineers to innovate together in a way that would have a real impact for customers. While the case wasn't a certain success, Tom was optimistic and recommended that we take immediate action in federal court in Virginia on Tuesday morning. I gave the approval they needed.

One aspect of this novel approach was an easy win—we were pretty much guaranteed the hackers would not show up in court to defend themselves. How could they? They would subject themselves to jurisdiction and even prosecution. The DCU team had succeeded in achieving something that we'd always sought to do but that was typically difficult to achieve. Our legal strategy had turned the hackers' strength—their ability to hide in the shadows—into their weakness.

We won the court case, took control of the internet domains, and started to contact and work with the victims. While the court documents were public and one security publication reported on what we'd done,[3] the rest of the press paid no attention. We felt confident expanding the tactic. We returned to court fourteen times, seizing ninety Strontium domains, and we persuaded the court to appoint a retired judge as a special master who could approve our applications more quickly.

By early 2017, we had uncovered activity aimed at hacking into the campaign teams of candidates for the French presidency. We alerted campaign staff, as well as the French national security agency, so they could put stronger security measures in place. We used our data analysis capabilities to identify current and evolving trends, and we developed an AI algorithm to predict the domain names hackers would go after in the future. But none of this was a panacea. It was just another round of cat and mouse. But at least the cat had some new claws.

Unfortunately, the mouse was becoming more sophisticated in ways that no one completely understood during the US presidential race. In 2016, the Russians had weaponized email to harvest and leak stolen communications, publicly embarrassing the leaders of Hillary Clinton's campaign and the Democratic National Committee with its contents.[4]

In 2017, they took their scheme one step further in France, leaking a combination of genuine and falsified emails attributed to Emmanuel Macron's presidential campaign.[5] While the Microsoft DCU and teams across the tech industry found new ways to thwart the problem, it was soon apparent that the Russians were innovating as quickly as we were.

We followed Strontium as it hacked its way around the world. Remarkably, the trail led to targets in more than ninety countries, with the highest amounts of activity in central and eastern Europe, Iraq, Israel, and South Korea.

In ordinary times, there would have been a strong and unified response from the United States and its NATO allies. But this was no ordinary time. Because the incidents in the United States became so intricately bound with the perceived legitimacy of the 2016 presidential election, all potential bipartisan discussion went out the window. At a meeting with a bipartisan group of our political consultants in Washington, DC, I argued that both parties were letting us down. Many Republicans were reluctant to take on the Russians because they thought it implicitly meant undercutting a Republican president. And some Democrats seemed to find more joy in criticizing Donald Trump than in taking effective action to address the Russian government. As a result, a principal pillar of the post–World War II defense of democracy had crumbled before our eyes—a united and bipartisan American public supporting US leadership that could bring together our NATO allies. As I shared my frustration, our consultants nodded when one said, "Welcome to Washington."

It seemed unlikely that the tech sector would turn this tide by itself. In late 2017, I heard an appeal directly from government officials while visiting Spain and Portugal, where there was mounting concern about Russian hacking. While there was pressure and clear recognition that we needed to do more, it was difficult to rally public support without explicit and more pointed discussions about what we saw happening.

One of our biggest challenges was how to talk publicly about the threats. Every tech leader was reluctant to name names, and we were no different. We were companies, not governments, and while we all had lived through governmental criticism before, we weren't accustomed to accusing a foreign government of misusing our platforms and services. But it was becoming increasingly apparent that our silence risked further enabling the very threats we wanted to help stop.

We wrestled with the "Russian problem" internally. If we publicly discussed the Russian government's connection to hacking, we worried it would retaliate against our business interests and employees there. We tried to reassure our private- and public-sector customers in Russia that our concerns with their government didn't mean that we would turn our backs on them or their country. After all, we had sued our own government five times, under Obama and Trump. We had not minced words with the Trump administration when it came to immigration issues. It didn't mean we weren't committed to continuing engagement and support across the United States, but how could people around the world expect us to be critical of US steps against surveillance and immigrants but not Russian steps against democratic societies?

In late 2017, we spotted renewed email-hacking activity on our services appearing to target Senate incumbents up for reelection in the 2018 US midterms. We alerted the Senate offices being targeted before any account was compromised. No one wanted to publicly comment on the disrupted attacks, so we remained silent.

In July 2018, Tom Burt spoke at the Aspen Security Forum and mentioned on a panel that we had observed and helped disrupt two phishing attacks against members of Congress seeking reelection. He didn't disclose the names of the members, and the press paid relatively little attention. But the tech news website *The Daily Beast* did some research and identified accurately Missouri senator Claire McCaskill as one of the two members.[6] Suddenly press interest exploded, and we soon heard about briefings taking place in the White House situation room to discuss the situation. McCaskill quickly did what we had hoped she would have done when we first put the request to her staff: She issued a strong statement, saying, "While this attack was not successful, it is outrageous that they think they can get away with this. I will not be intimidated."[7]

We learned an important lesson. Congressional staff weren't any more accustomed than we were to publicly discussing these types of attacks. Especially if we started with the IT staff for an organization, the decision making would spin in circles for months, which effectively meant that no one would say anything at all. But if the question was called at the top of an organization, it wasn't all that difficult for people to know what they wanted to say.

While talking more publicly was important, as the provider of these online services, we realized that we also needed to do more. We decided to develop a specific program to better shield political candidates, campaigns, and associated groups from online meddling. We called the program AccountGuard. The service would be offered free of charge for the political groups and individuals using our Office 365 email and services. MSTIC would actively monitor nation-state activity and we would alert campaign staff with detailed information when an attack was detected.[8]

I loved the initiative, but we knew AccountGuard was just part of the answer. If democratic leaders around the world were going to step up more strongly to defend against expanding election meddling, the time had come for the tech sector to get more explicit about what we were seeing.

The AccountGuard announcement provided an opportunity to break through. We had recently watched Strontium as it had created six websites that clearly targeted American politicians. Three were focused on the US Senate, and two others were particularly noteworthy. One of these appeared to target the International Republican Institute, or IRI, which was a leading Republican organization that supported democratic principles around the world. The other appeared to target the Hudson Institute, a conservative think tank that had objected strongly to a variety of Russian policies and tactics. Put together, these provided a solid indication that Strontium was not targeting Democrats alone and instead was focusing on both sides of the American political aisle.

The DCU secured a court order to transfer control of all six sites to our sinkhole. We concluded that we had acted before anyone had been hacked. Now the question was how public we should be about it. There was sure to be an active debate among the various groups within Microsoft. But it was a good time to encourage a more wide-ranging public discussion, especially now that the hacking was impacting both political parties.

This launched a week of vibrant internal debate that concluded with a call in my office on a Friday morning. We decided to contact the heads of the two private organizations and officials at the Senate and let them know in advance about our plans to go forward with an announcement the following Tuesday.

Leaders of both organizations were quick to support our actions. As one put it, the attacks in some ways were "a badge of honor" that recognized the importance of what they were doing. We coupled our AccountGuard announcement with information about these new attacks and an explicit statement that the six websites had been created by "a group widely associated with the Russian government and known as Strontium, or alternatively Fancy Bear or APT28."[9] It marked the first time we had been this explicit in identifying Russia as the source of the attacks, a move that was followed within days by both Facebook and Google as they acted to take down disinformation and fake accounts from their sites.

While this was hardly the end of our journey, it marked how far the tech sector had come since 2016. As the industry took new steps, the press began to urge the US government to match our efforts. It provided what we hoped would become a new foundation for broader and more collaborative action. As I made the case on *PBS NewsHour*, we needed to "set aside enough of our differences to work together to do what it takes to secure our democracy from these kinds of threats."[10]

Perhaps not surprisingly, the Russian government was less than enthusiastic about the tech sector's stronger public stance. In November 2018, a Microsoft employee in Redmond applied for a visa to attend an AI conference in Moscow. He was summoned more than two thousand miles to the Russian embassy in Washington, DC, for a "visa interview." As he walked into the interview room, a consular official handed him an envelope and politely asked him to read the two documents inside. The official then asked our employee to take the documents back to Redmond to hand them over to Microsoft executives. The interview ended less than five minutes after it had begun, and the employee's visa was granted.

I soon received an email with the two documents attached. They were printed copies of English-language versions of official Russian news stories. Both were reports detailing my statements made that August, noting the Russian government's disagreement with my characterizations. As one report concluded, "The Russian authorities have repeatedly refuted any accusations of interference in elections abroad, including through hacker attacks."[11]

Russia's message to Microsoft reflected the bind that many American tech firms now confront. On the one hand, US politicians understandably push us to take strong stances against foreign hacking. But on the other, these steps lead to foreign pressure on the companies themselves.

As the full picture of Russian activities began to take shape, it was apparent that emails were not the only digital technology that risked being weaponized. One of the important lessons in the field of risk management is that you need to think about both the risk that you're most likely to face and the risk that, even if unlikely, would be the worst to have to face. As we consider the digital risks to democracy, it's hard to imagine anything worse than that of potential hacking of voting machines or the disruption of accurate electoral tallies. Imagine the impact if, after a close and important election, news emerged that a foreign government had hacked our voting systems in significant ways that can't be rectified. To borrow Franklin's phrase, how can we "keep a republic" if the public loses confidence in whether the votes being reported have actually been cast?

Already the world has witnessed nation-state probes to assess whether they can tamper with voting machines. And academics have documented the vulnerabilities in many voting machines, which have computer software and hardware that were developed in the early 2000s. While public funding to tackle the problem is increasing, more action is needed to address what is an obvious and long-documented vulnerability in these aging computer systems.

It's a problem the tech sector needs to help address. Innovative efforts are starting to spread, including at Microsoft, where research led in May 2019 to the launch of ElectionGuard, an encrypted voting system that protects individual ballots and their collective tallies.[12] It's an open-source-based software system that uses inexpensive and off-the-shelf hardware and combines the best of old and new technology. A voter chooses candidates on an electronic screen, which then records these choices on a paper ballot that's printed and the voter deposits, ensuring a paper record for any post-election audit that might be needed. The voter also receives a personal printout with an electronic tracking number that uses an encrypted algorithm to reflect his or her selections. This tracking record can be checked online later to confirm that an individual's votes remain recorded accurately. The overall solution provides a trusted and secure count of every vote cast. It's the type of approach that is vital to the security and functioning of a democracy.

Cyber-based threats around hacking of campaigns and disruption of voting were barely on the radar screen a decade ago. Today they are real risks that spill into daily news reports. Just as democratic governments and industry worked together to win a world war in the 1940s, today they must develop a unified response to protect the peace.

And as authoritarian regimes experiment with disinformation campaigns, even more complex challenges lie ahead.

Chapter 6 ⟫ # SOCIAL MEDIA:
The Freedom That Drives
Us Apart

In a museum in the center of Tallinn, Estonia, which sits on the edge of the Baltic Sea, a young woman and man spin in perpetual motion, perched on opposite ends of a long, narrow plank. With arms outstretched and locked gazes, they steady themselves and each other while a giant seesaw slowly rotates on a narrow fulcrum. While whimsical, the curious sculpture sends an unmistakably serious message.[1] It represents the fragile balance that free societies around the world now confront: protecting democracy in a social media age from the freedoms that can drive people apart.

The spinning sculpture serves as a final page in the museum's narration of a story tracing almost a century of the Baltic nation winning, losing, and rewinning sovereignty, and the hard work required to retain it. It's also a story that speaks to the technology challenges confronting every modern democracy. As the audio guide tells visitors, "Estonia didn't become free in one day. We're seeking freedom. We do it every day."

The two-story Vabamu Museum of Occupations and Freedom sits down the hill from Tallinn's medieval city center. While modest in size, the glass and steel building stands in proud contrast to the walled thirteenth-century capital's fortified town center that soars above it. Built as an emblem of Estonia's new era, walls of windows beckon the northern light into the contemporary building, illuminating a modern stage that tells a complex and sad tale written by Russian, Nazi German, and Soviet occupiers. But the museum doesn't focus solely on hardship, oppression, and murder. It sounds a common chorus of people around the world yearning for freedom. And most important, it examines the constant tension between freedom and responsibility that the museum's pair of floating mannequins so elegantly display.

When we visited Estonia in the fall of 2018, the US Congress's investigation into disinformation campaigns on Twitter and Facebook was at full throttle. The world had awoken to this new set of challenges and was asking questions. How had this happened? Why had it happened? And why hadn't we realized it sooner?

One answer to these questions came to us on a Saturday morning at the Vabamu Museum, the brainchild of an Estonian turned American named Olga Kistler-Ritso. Born in 1920 in Kiev, Ukraine, as the Russian Empire collapsed, Olga came of age under a series of authoritarian regimes. As a young girl, she and her older brother escaped the tumult and famine of Ukraine by emigrating north to Estonia. Toward the end of the Second World War, as Soviet troops prepared to sweep the tiny country back into the fold, Olga, then a young woman, fled with retreating German soldiers on one of the last ships to leave the country.

In 1949, Olga came to the United States, eventually settling with her husband and daughter just minutes from what would become Microsoft's corporate headquarters in Redmond, Washington.

While she lived for the rest of her life in the United States, Estonia never left her thoughts. She kept tabs on her childhood home, which continued under Soviet occupation.[2] That changed in 1991 when, after more than fifty years of occupation, Estonia broke free from Russian rule and began building its future as an independent nation.

Eager to contribute to Estonia's democratic future, Olga donated her life's savings to build a museum that would memorialize an important story that she wanted to ensure the world would neither forget nor repeat. As President Lennart Meri, the institution's patron, said at its opening in 2003, the building is much more than a museum. "This is the house of freedom and it should constantly remind us how subtle and sensitive the barrier is between freedom and its opposite, totalitarianism."[3]

Each year, the Vabamu Museum takes more than fifty thousand visitors from around the world on Estonia's journey through occupation and freedom—and, as it turns out, how technology can become a weapon.

The internet helped push Estonia out of the shadows of Communism, transforming what would become the home of Skype into a vibrant self-proclaimed "e-democracy." But in 2007, Estonia's former occupier struck the country's digital underbelly, revealing democracy's inherent fragility and how the very technology that helped define a nation's freedom also made it more vulnerable.

That spring, Estonia suffered the first cyber-based nation-state attack on another country, a digital siege called a denial-of-service attack that froze much of the country's internet, including sites that powered Estonia's government services and economy. The world suspected Russia.[4]

"If something barks like a dog, it's a dog," said Marina Kaljurand, Estonia's former foreign minister, during our lunch in Tallinn. "But in our case? It was a bear!" Marina should know. At the time of the attack, she was Estonia's ambassador to the Russian Federation.

The 2007 attack put the country of 1.3 million people on the cybersecurity map. As a result, NATO built its Cooperative Cyber Defense Centre of Excellence just outside Tallinn. Living in Russia's crosshairs forced the country and its leaders to focus not just on war and peace but also on freedom and oppression, issues that occupy opposing sides of the current information technology equation.

The museum that Olga built shows the collision of technology and society in a way that few places can. People who are oppressed are united by a common desire, the quest for freedom. But once people are free, that common bond fades away. The people of Estonia learned firsthand after the Iron Curtain came crashing down that freedom brings its own challenges, and they can be dizzying.

"In some sense that's actually a very frightening situation because everybody has trouble figuring out what they really want," the exhibit states. "So what should you want if everything's allowed? And then people run themselves ragged in all directions."

Facebook CEO Mark Zuckerberg created his online platform to make the world more "open and connected." At one level, it's the ultimate endorsement of freedom. But in a country where KGB intelligence agents had registered, tracked, and pulled printed samples from every typewriter in the country to discourage unsanctioned communications, Estonians know all too well how overwhelming it can be when the tap of information and ideas suddenly flows freely.

So what do people do? As the museum exhibit observes, they find their tribe—in this case, their cybertribe. People seek out online groups of like-minded people that replicate communities that have always characterized human society. These groups in turn become more connected but less open, choosing their preferred channel and the people they want to interact with. They share information based only on a single vantage point. As in the real world, people can be quick to believe the worst about others, especially people they perceive as different from them. People's defense mechanisms start to kick in. Idealism, in short, collides with human nature.

Who figured this out and capitalized on it before anyone else? It was people who, like the Estonians, had lived their lives under a combination of repression and freedom, perhaps able to appreciate this dynamic more quickly than others—the Estonians' neighbors across the Russian border. And who were the last to wake up? Idealistic Americans on the west coast of the United States who had lived their entire lives in freedom.

But to truly understand this phenomenon, we need to recall another consequence of technology that reinforces our tendency to splinter into cybertribes: being alone together.

More and more, we find ourselves absorbed in electronic conversations with people we aren't physically with. Sometimes we are a world apart. Digital technology has made the world smaller—and people more accessible—but it has also cast a deafening silence between people sitting next to each other. This phenomenon is nothing new. For more than a century, almost every technology that has connected people who live apart has also created new barriers between people who live close together.

No other modern technology has reshaped our lives quite like the automobile—and in few places more profoundly than in rural America. Until the start of the twentieth century, rural dwellers typically shopped, worked, worshipped, learned, and socialized within the twenty-mile radius that their horse and buggy carried them. The general store was the town hub, children of all ages attended a one- or two-room school, and a small village church served the entire community.

All this changed once gas-powered cars hit the countryside.

Between 1911 and 1920, cars on farms alone swelled from eighty-five thousand to more than one million.[5] The automobile and modern roads opened up new vistas with opportunities farther afield, narrowing the urban-rural divide. As one historian has noted, the car freed "rural people from the physical and cultural isolation that was a characteristic feature of life in the countryside."[6]

But there was also a cost for this greater mobility.[7] The more time people spent somewhere else, the less time they spent with their families and neighbors. The car frayed forever the tightly knit fabric of small towns.

Beginning in the 1960s, landline telephones had a similar effect on families. For teenagers, spending time alone in their bedrooms now meant spending time with their friends on the phone and later on their computers. The members of a family found themselves alone together in the same home.

Forty years later, smartphones would bring kids back into closer physical proximity with their parents, but their minds were clearly somewhere else. And it became commonplace for families to argue about putting those phones down, especially at the dinner table. Repeatedly over time, technology has made the world a smaller place, but people are less connected with those living next door or under the same roof.[8]

This also creates new challenges for democracy. Spending more time online, sometimes with complete strangers, has made people more susceptible to disinformation campaigns that play to their likes, desires, and sometimes their prejudices, with real-world consequences.

For decades one of the strengths of the world's republics had been the ability to use open communication and public discussion to ensure broad and even bipartisan understanding, support for foreign policy issues, and a commitment to democratic freedoms. It was seldom an easy task, but as Franklin Roosevelt proved, new communication technologies, like the radio in his time, could be used to build public support for difficult measures such as supporting the United Kingdom before the United States entered World War II. And the United States used everything from radio to the fax machine to spread information and nurture democracy in the closed societies in Central and Eastern Europe in the decades that followed.

But now others have turned the tables on this very strength of a free and open society. Hacked emails might be the tip of Russia's new spear, but their ambitions have a far broader reach. Cable news and then social media have created increasingly separate information bubbles in Western democracies, especially in the United States. What if information—true or otherwise—could be spread using platforms like Facebook and Twitter to rile up various factions and undermine political candidates likely to be more hostile to Russia's interests? What if teams of technologists and social scientists joined forces in Saint Petersburg and Moscow to influence the American political and social narrative with the same creativity and speed as the creators of the platforms they were exploiting? And what if no one in the United States was paying enough attention to see what was happening?

In late 2018, a team from Oxford University and the American analytics firm Graphika analyzed subpoenaed data that Facebook, Instagram, Twitter, and YouTube provided to the Senate Intelligence Committee. The team was the first to document thoroughly how Russia's Internet Research Agency (IRA) had "launched an extended attack on the United States by using computational propaganda to misinform and polarize US voters."[9] These disinformation efforts typically peaked near key dates on the American political calendar, a strategy that played into the interactive and viral nature of social media platforms. As the report found, between 2015 and 2017 more than thirty million users "shared the IRA's Facebook and Instagram posts with their family and friends, liking, reacting to, and commenting on them along the way."[10]

By manipulating American-made technology, the Russians were able to reach into and stir the US political pot. This foreign influence spilled over into the real world, notably during the IRA's successful effort in 2016 to organize a synchronized protest and counterprotest in Houston.[11] Neighbors shouted at neighbors, unknowingly egged on by people in Saint Petersburg, Russia.

By late 2017, that reality had become increasingly clear. Still, when reports of Russian disinformation efforts on Facebook emerged, most people in the tech sector, including Mark Zuckerberg, were skeptical that the activity was widespread or that it could have much effect.[12] But that soon changed. By the fall of 2017, Facebook found itself in the crosshairs of officials around the world. The social media giant was under more public scrutiny than any tech company since the antitrust cases against Microsoft almost two decades earlier. Having lived through those years at Microsoft, I appreciated the important and rising government demands on Facebook. I also understood the enormous difficulties the company faced. Facebook had not designed its services as a platform for foreign governments to use to disrupt democracy, but neither had it put in place measures that could prevent or even recognize such activity. No one at the company—or in the tech sector or the US government, for that matter—had anticipated such a phenomenon until Russia turned Facebook against the very country that had given it life.

I was especially struck by the world's focus on Facebook when we attended the Munich Security Conference in February 2018. Founded in 1963 and now led by respected former German diplomat Wolfgang Ischinger, the annual summit brings together defense ministers and other military and national leaders from around the world to discuss international security policy. In 2018, the attendee list included some of my peers from the information technology industry.

As I made my way through the wall of high-ranking military service officers in the packed lobby of the Bayerischer Hof hotel, I felt a bit out of place. It was a homecoming of sorts as I squeezed into the elevator next to Eric Schmidt, then chairman of Google, and his team. It was an odd place to run into someone from Silicon Valley.

"Have you been here before?" he asked.

"Actually, it had never really occurred to me before that I needed to be here," I replied.

But times had changed and in 2018, it was important that we were both in Munich.

Much of the discussion that week focused on the weaponization of information technology. At a lunch with CEOs, International Monetary Fund head Christine Lagarde was asked why she had come to a defense conference. She explained that she wanted to understand how information technology was being used to harm democratic processes to help her think about how it might be misused to attack financial markets. It was a sobering conversation, but I was reassured by her foresight.

While the conversations were difficult and a bit heavy, I couldn't help but have a bit of sympathy for Facebook's chief security officer, Alex Stamos, who was on the defensive throughout the entire conference. During a panel we sat on together, he was peppered with sharp questions by a rising Dutch member of the European Parliament. Later that evening during dinner with the Atlantic Council, government officials and other indignant attendees challenged him repeatedly, asking how Facebook "had allowed all this to happen."

While the concerns were understandable, I was increasingly exasperated by the discussion. Everyone was pointing a finger at Facebook, but no one was pointing a finger at the prime culprit. It was like yelling at the person who forgot to lock the door without talking about the thief who broke in.

The bigger question for Facebook, the United States, the world's democratic republics, and the entire tech sector was what to do. The reaction of some in government was to cast blame on Facebook and other social media companies, insisting that they solve the problem. While the companies who invented this technology indeed held much of the responsibility, such an approach seemed incomplete. The answer would require a mix of action by governments and by the tech sector itself.

In the summer of 2018 as Mark Zuckerberg testified before Congress, the tech sector had changed its tune on the magnitude of the problem and the response it required. "My position is not that there should be no regulation," Zuckerberg said. "I think the real question, as the internet becomes more important in people's lives, is what is the right regulation, not whether there should be or not."[13]

As his statement reflected, it's one thing to recognize the obvious and acknowledge that regulation is needed. It's quite another to figure out what type of social media regulation makes sense.

One person leading the charge to answer the latter question is a former telecommunications executive serving in the United States Senate since 2009, Mark Warner from Virginia. In the summer of 2018, Warner released a white paper with a series of proposals designed in part to address disinformation campaigns through new legislation.[14] He acknowledged some of the technical and privacy issues associated with these issues and called for more discussion.

As Warner reflected in his paper, an emerging issue for social media on the internet is the growing disquiet with its current immunity under the US Communications Decency Act. Congress in 1996 passed legislation to nurture the internet's growth in part by shielding the publishers of "interactive computer services" from many of the legal responsibilities that more traditional publishers face. For example, unlike television and radio, social media services bear no legal responsibility in the United States under state and many federal laws for illegal content published on their sites.[15]

But the internet is no longer in its infancy, and its impact today is globally ubiquitous. As nation-states, terrorists, and criminals exploit social media sites for nefarious purposes, political leaders increasingly are joining traditional publishers in questioning whether social media sites should continue to get a legal pass. Warner points to the expected spread of "deep fakes," or "sophisticated audio and image synthesis tools that can generate fake audio or video files falsely depicting someone saying or doing something," as an additional reason to impose new legal responsibilities on social media sites to police their content.[16]

As the world has witnessed more horrifying acts amplified on social media, political pressure has grown. A decade from now, we may look back at March 2019 as an inflection point. As Kevin Roose wrote in the *New York Times*, the horrific terrorist slaying of fifty-one innocent Muslims on March 15 in two mosques in Christchurch, New Zealand, in some ways "felt like a first—an internet-native mass shooting, conceived and produced entirely within the irony-soaked discourse of modern extremism."[17] As he described, "The attack was teased on Twitter, announced on the online message board 8chan and broadcast live on Facebook. The footage was then replayed endlessly on YouTube, Twitter, and Reddit, as the platforms scrambled to take down the clips nearly as fast as new copies popped up to replace them."[18]

Just two weeks later we were in Wellington, New Zealand's capital, on a trip that had been months in the making. New Zealand's Prime Minister, Jacinda Ardern, who had handled the shock and crisis with extraordinary judgment and grace, had given a speech that captured a marked shift toward social media. "We cannot simply sit back and accept that these platforms just exist and that what is said on them is not the responsibility of the place where they are published," she said.[19] She then referred to social media sites even more emphatically, "They are the publisher, not just the postman. It cannot be a case of all profit, no responsibility."[20]

When we met with Ardern and her cabinet members in New Zealand, I could not disagree. The episode demonstrated that tech companies needed to do more, including Microsoft's own services such as Bing, Xbox Live, GitHub, and LinkedIn. And, more broadly, a regulatory regime established nearly a quarter century ago suddenly seemed insufficient to address the threats to the public from hostile nations and terrorists alike.

While there are clear distinctions between the exploitation of social media platforms by terrorists and state-sponsored attackers, there are similarities. Both involve intentional efforts to undermine the social stability on which societies depend. And, as it turns out, politically the responses to both problems may reinforce each other, pushing governments to move toward a new regulatory model for social media sites.

The move to regulate social media may seem unprecedented, but it's worth recalling that the United States has experienced some of this before. Much of what we are seeing parallels efforts to regulate radio content in the 1940s.

The first wireless radio program aired in the United States in November 1920 by Westinghouse, broadcasting William Harding's victory in the presidential race to succeed Woodrow Wilson.[21] When it first came into people's homes, it was considered a modern marvel. It connected the world through common experiences, beaming live events, entertainment, and breaking news. The wireless rocketed to popularity in the 1930s and became a fixture in 83 percent of American living rooms by the end of the decade.[22] It was the Golden Age of radio, and the technology shaped everything from American culture and politics to family life.[23]

As radios became ubiquitous in the latter half of the 1930s, concerns about its societal impact spread. As noted in a 2010 article in *Slate*, "The wireless was accused of distracting children from reading and diminishing performance in school, both of which were now considered to be appropriate and wholesome. In 1936, the music magazine *The Gramophone* reported that children had 'developed the habit of dividing attention between the humdrum preparation of their school assignments and the compelling excitement of the loudspeaker' and described how the radio programs were disturbing the balance of their excitable minds."[24]

After World War II ended, there emerged what scholar Vincent Pickard has termed "the revolt against radio."[25] As he documents, while the radio market initially grew on a business model that provided free programs as a lure to sell radio receivers, by the 1940s, most American homes already had one or more radios inside them. The business model for radio programs evolved towards advertising, which in the eyes (or, more accurately, the ears) of some critics, led to soap operas and other programs that became increasingly meaningless and even vulgar. As Pickard observes, "This criticism took shape across grassroots social movements, commentary from varied newspapers and opinion journals, as well as hundreds of letters from average listeners to editors, broadcasters, and the FCC."[26]

Impatience reached a peak, leading the Federal Communications Commission in 1946 to publish its Blue Book, a report named after its blue cover that sought to make "the privilege of holding broadcast licenses contingent upon meeting substantive public interest requirements."[27] Commercial broadcasters unleashed a political back-lash against the report and defeated its proposals, but the episode nonetheless altered broadcasting history, causing the major radio networks to fund documentaries and improve their public interest programming.[28]

Some might look at the revolt against radio and find in this history reason to believe that a challenge to social media might similarly represent a passing political moment, unlikely to lead to lasting regulatory change. While there's never certainty when predicting the future, there's reason to believe that, to the contrary, the issues concerning social media will be more, rather than less, impactful. One reason is that current concerns such as nation-state disinformation and terrorist propaganda speak to more serious issues than the debate about banal programming in the 1940s. And a second reason is the global nature of current regulatory proposals. While the United States has traditionally been reluctant to regulate content, given the importance of the First Amendment to the Constitution, among other factors, other nations do not protect freedom of speech to the same degree.

If there was any doubt about this latter aspect, it was quickly laid to rest by events in Australia that followed the attack in Christchurch, New Zealand. Within less than a month, the Australian government passed a new law requiring social media and similar sites to "expeditiously" remove "abhorrent violent material" or risk criminal penalties that can include up to three years in prison for tech executives and a fine of up to ten percent of a company's annual revenue.[29] While many across the tech sector responded with angst to what they felt was a combination of strong criminal penalties and imprecise legal standards, the development spoke to the mounting frustration of political leaders around the world, making clear the political demands for replacing online services' legal immunity with a new regulatory model.[30]

There remains a big difference, however, between needing something new and knowing precisely what's needed. It seems impossible for social media sites to follow the pre-publication editorial review processes that are used by traditional print, radio, or television outlets. Imagine if every photo on Facebook or entry on LinkedIn needed to be reviewed by a human editor before it could be viewed by others. It would "break the mold" that makes it possible for hundreds of millions or even billions of users around the world to upload content and share it with family, friends, and colleagues.

This is a problem that needs to be solved with a scalpel rather than a meat cleaver. It's not an easy challenge, especially in moments of political pressure. It was in part to avoid a hasty legislative reaction that in 2018 Warner encouraged a conversation with social media platforms—only to receive little or no feedback from some of the most prominent companies. Worrying about mounting Russian exploitation of social media, he offered a menu of more tailored approaches. One of his ideas, which was taken further by the Australians, is to obligate social media platforms to prevent users from continuing to re-upload illegal content, effectively increasing the legal responsibility to act once a problem has been verified.[31] A more general variant was proposed by the British government just two weeks after the Australians acted, recommending a new "statutory duty of care to make companies take more responsibility for the safety of their users" and backing this with oversight by an independent regulator.[32] Warner also proposed rules that would impose a duty on social media platforms to determine the origin of accounts or posts, identify bogus accounts, and notify users when bots are spreading information.

As these show, there is likely to be room for complementary regulatory approaches, combining a more narrow focus on specific categories of objectionable content with a broader effort to provide users with more information on its sources. One important feature of the latter approach is its emphasis on addressing the spread of disinformation not by assessing whether content itself is true or false but instead by providing social media users with accurate information on people's identity. It's a commonsense approach that's been adopted in modern-day political advertising. Leave it to the public to decide what is true. But let them make that decision based on an accurate understanding of who is speaking. And in the twenty-first century, let the public know whether it's a human being or an automated bot that's doing the talking.

Interestingly, the same approach is used by a nongovernment initiative launched by two prominent Americans from the media sector—one a conservative and the other a liberal. Gordon Crovitz is the former publisher of the *Wall Street Journal* and Steven Brill is a former journalist who founded *The American Lawyer* and Court TV. Together they created NewsGuard, a service that relies on journalists to create what they call "nutrition labels" for the media.

Working through a free internet browser plug-in, NewsGuard displays green or red icons next to links on search engines and social media feeds, including Facebook, Twitter, Google, and Bing, indicating whether a site is "trying to get it right or instead has a hidden agenda or knowingly publishes falsehoods or propaganda."[33] In addition to rating news and information websites, NewsGuard applies a blue icon for platform sites that host user-generated content and an orange icon for humor or satire sites designed to look like legitimate news. There are gray icons for websites that haven't yet been reviewed and rated.[34]

The effort hasn't been without teething challenges, especially as the NewsGuard team has expanded outside the United States and sought to develop ratings criteria that are effective worldwide. But Crovitz and Brill can move a lot faster than a government. Their service was up and running before Warner's proposals reached a congressional hearing, and the team continues to refine and improve its operations. And as a nongovernmental effort, it can quickly expand internationally. But it depends on private funding and on tech companies for browser plug-in support and ultimately on users themselves to adopt the service.

Ultimately, there are two broader lessons that emerge. First, initiatives from the public and private sectors will likely need to move forward together and complement each other. And, second, despite the novelty of current technology, there is a lot to learn from the challenges of the past.

Interestingly, foreign interference in democracy is almost as old as the United States itself. A democratic republic by its very nature is subject to disruption—both foreign and domestic—by efforts to disrupt confidence and sway public opinion. The first person to realize this was an early French ambassador to the United States named Edmond Charles Genêt. He arrived in America in early April 1793, just a few weeks before President George Washington officially declared the United States' neutrality in the expanding war between France and the United Kingdom. Genêt was on a mission to tip the young republic toward supporting France, including by persuading the United States to accelerate the repayments of its debt to his country and by enabling attacks on British shipping by armed privateers operating from US ports. If required, he was prepared to spark an attempt to overthrow the country's young government.

Genêt's arrival sparked increasing tension within Washington's cabinet, with Thomas Jefferson sympathetic to the French and Alexander Hamilton sympathetic to the British. Genêt sought to appeal directly to the American public for his cause, a move that, in the words of one historian, did more than spark the origins of our two-party system. "Political dialogue was impassioned, street brawls were not uncommon, and old friendships were severed."[35] In 1793, Washington and the members of his cabinet overcame their differences and united in demanding Genêt's recall to France.[36]

That outcome has a lesson for our own generation. Foreign interference with democratic processes can be met successfully only if the stakeholders in a republic set aside enough of their differences to work together to respond effectively. It may be difficult to remember that the differences between Jefferson and Hamilton and their respective supporters were as passionate as the disagreements between Republicans and Democrats today. But the Broadway musical *Hamilton* provides a powerful reminder that at least today our politicians no longer resort to armed duels. The reality is that passionate divisions and even vitriol are an inherent risk and constant challenge for any democratic republic.

It was against this backdrop and continuing French attempts to tamper with American politics that Washington used his farewell address in 1796 to warn against the risks of foreign influence. "A free people," he said, "ought to be constantly awake, since history and experience prove that foreign influence is one of the most baneful foes of republican government."[37] Historians sometimes discuss the implications of his address when weighing the pros and cons of international involvement in the world. But we would do well to remember that Washington also focused more on the conflict at hand and direct foreign involvement in American politics, dealing firsthand with the risks this created.

Of course, many things have changed in the centuries since Washington uttered these words. In his time, those seeking to influence public opinion turned to newspapers, pamphlets, and books. Then came the telegraph, radio, television, and the internet. Today, someone in a cubicle in Saint Petersburg can respond within minutes to a political development anywhere in the world with targeted disinformation.

The United States government itself has used information technology to inform and even persuade the public in other countries to support certain stances. Some of this has been clandestine; many in the United States today would reject some of the steps taken by the CIA in Europe and Latin America in the 1950s. But others have been out in the open, including Radio Free Europe during the Cold War and today's Voice of America.

The United States as a nation has been comfortable using technology to spread information to seed and advance democracy. But now technology is being used to spread disinformation and disrupt democracy. At one level, we can divide these activities into separate categories based on principles that we connect with fundamental human rights. But at another level, realpolitik has changed in a critical respect. Until recently, communications technologies seemed to favor democracy and put authoritarianism on the defense. Now we must ask whether the internet has created an asymmetric technology risk for democracies that authoritarian governments can counteract more readily than the republican form of government that Franklin's words urge us to protect.

The answer is probably yes. Digital technology has created a different world, and not always a better one. And how we address this is not yet completely clear. But as in Washington's time, it will require that the stakeholders in democratic republics work together, not just across political parties, but across the tech sector and with governments around the world.

Chapter 7

DIGITAL DIPLOMACY:
The Geopolitics of Technology

When Casper Klynge arrived on Microsoft's Redmond campus in February 2018, he could have been mistaken for a tech entrepreneur. Or given his sharp dress, California vibe, and less-than-close shave, perhaps an actor or musician. When I shook his hand, I paused to recall whom I was meeting.

Casper is not your typical ambassador. And he doesn't have a typical assignment. He is the first person to serve as Denmark's tech ambassador, responsible for connecting the Danish government to tech companies around the world. His "embassy" has more than twenty employees working on three continents, with staff in the United States, China, and Denmark.

When I had met with a group of European ambassadors in Copenhagen the preceding spring, Casper's new job was on people's minds. The Danish foreign minister, Anders Samuelsen, had proclaimed the position "a world first" and a necessity, stating that tech companies affect Denmark as much as countries do. "These companies have become a type of new nation and we need to confront that."[1]

While Denmark was the first country to name a formal ambassador to liaise with the tech sector, the country's decision followed a similar step by the British government. In 2014, Prime Minister David Cameron created a position in his office for a special diplomatic role, initially to address law enforcement technology issues and then to serve as "special envoy to US technology companies." The first to take the new post was Sir Nigel Sheinwald, a former British ambassador to the United States.

Other governments from Australia to France have followed suit with similar moves. It's a shift that shows how the world has changed.

Large corporations have played a major role in economies and societies since the birth of the business empires of the Gilded Age. And no industry transformed American society and ultimately American law like the railroads in the second half of the 1800s. As *Poor's Manual of the Railroads of the United States* so aptly put it at the turn of the century, "No enterprise is so seductive as a railroad for the influence it exerts, the power it gives, and the hope of gain it offers."[2]

Railroads were America's first big business, crossing state lines with thousands of miles of track, and they sparked a surge of regulation and laws governing commerce, patents, property, and labor. James Ely's *Railroads and American Law* may seem an unlikely book to see on a software executive's shelf, but it's one I refer to periodically to help me think about how technology changes the world around it.[3]

While the railroad may be considered by some as the internet of its day, there's something quite different about today's digital technology. The products and companies are far more global, and the pervasive nature of information and communications technology increasingly thrusts the tech sector into the center of foreign policy issues.

In 2016, a mantra, "There's no national security without cybersecurity,"[4] took hold within Microsoft and started to seep into the public discussion. We were hardly alone with this recognition. As German conglomerate Siemens AG predicted succinctly, "Cybersecurity is going to be the most important security issue of the future."[5] Clearly, any issue that would be fundamental to national security would propel the tech sector even more squarely into the world of international diplomacy.

In part, this made it more important to explain publicly and clearly what we were doing to address these issues. As we evolved our cybersecurity efforts, we recognized the need to take—and talk about—three distinct strategies. The first and most obvious was to strengthen technical defenses. This work naturally starts with the tech sector, but it becomes a shared responsibility when customers deploy these new services. At Microsoft we were spending more than $1 billion a year developing new security features, an investment that involved more than thirty-five hundred dedicated security professionals and engineers. This work is ongoing as we continually roll out new security features at an accelerating pace, and it's a huge priority across the tech sector.

The second approach, involving what we call operational security, was in some ways more of a priority at Microsoft than at some other tech companies. It includes the work of our threat intelligence teams to detect new threats, the focus of our Cyber Defense Operations Center to share this information with customers, and the work of the Digital Crimes Unit to disrupt and take action against cyberattacks.

The latter work brought us increasingly into an area more traditionally addressed by governments alone. And it raised some complicated questions. How should companies respond to specific attacks? Of course, we needed to help our customers recover from hacks, but how could we deter attacks in the first place? Was attacking back an option?

When this question was put to a group of tech leaders at a White

House meeting in 2016, the reaction was mixed. One attending executive was enthusiastic about empowering companies to attack back, but I worried that vigilante tech justice could lead to mistakes and even chaos. It's why I took comfort in the fact that we typically required our Digital Crimes Unit to solve issues in court, often in cases that also involved law enforcement. It grounded us in a legal system where public authorities played their proper role and we remained subject to them and the rule of law more generally. I felt there was good reason to stick with this approach.

With rising nationalism, including in the United States, global companies also needed an intellectual foundation to act in a global way. We challenged our peers to act as a "neutral digital Switzerland" committed to defending all our customers around the world, pledging to play 100 percent defense and zero percent offense. Every government, including those with more nationalistic points of view, should be able to trust technology. They too benefit when the tech sector pledges to protect all of our customers regardless of nationality and refrain from helping any government attack innocent civilians.

When we combined these two strategies, they still seemed an insufficient response to broadening attacks. We needed to prop the cybersecurity stool with a third important leg: stronger international rules and coordinated diplomatic action to restrain cyberthreats and help galvanize the international community to pressure governments to stop indiscriminate cyberattacks. Until there was a greater degree of global accountability, we worried that it was too easy for governments to deny any wrongdoing.

In January 2017, coincidentally the week before Denmark announced the creation of what would become Casper Klynge's job, we at Microsoft found ourselves discussing ways to help galvanize the tech sector and unite the international community around cybersecurity. I recalled that the International Committee of the Red Cross, or ICRC, had brought the world's governments together in 1949 to establish the Fourth Geneva Convention to better protect civilians in times of war. "Isn't it ironic that now we're seeing these attacks against civilians, and it's supposed to be a time of peace?"

Our public affairs leader, Dominic Carr, was quick to reply. "Maybe it's time for a Digital Geneva Convention," he said.

Bingo. Just as governments had pledged in 1949 to protect civilians in times of war, perhaps a Digital Geneva Convention could capture people's imagination about the need for governments to protect civilians on the internet in times of peace. It was an idea that could build on work already in motion by governments, diplomats, and tech experts focused on establishing so-called cybersecurity norms between nations. Perhaps a compelling example and brand would help us speak more effectively to the nontechnical audiences that we needed to win over if any of this were to become a reality.

We called for the continued strengthening of international rules to avoid cyberattacks that target private citizens or institutions or critical infrastructure in times of peace, as well as an expanded ban on the use of hacking to steal intellectual property. Similarly, we urged stronger rules to require governments to assist private sector efforts to detect, respond to, and recover from these types of attacks. Finally, we urged the creation of an independent organization that can investigate and share publicly evidence that attributes nation-state attacks to specific countries.[6]

After we shared our ideas at the annual RSA security conference in San Francisco in 2017, a number of journalists picked up the theme and enthusiastically zeroed in on the call for a Digital Geneva Convention.[7] While the press is always a good litmus test for the acceptance of new ideas, the bigger test was whether the conversation would change in national capitals. And a good way to gauge whether people were listening, ironically, is whether some disagree. After all, in a world with so many issues and such fragmented media, it was easy for many ideas to become like trees falling in the wilderness. Busy people in important positions couldn't possibly care enough to take the time to address most of what others said during a day.

We passed this test. In Washington, DC, those most bothered by the idea of a Digital Geneva Convention were often officials who had played a leading role in developing the nation's offensive cyber capabilities. They argued that rules restricting the use of cyber capabilities would hold back governments like the United States. We pointed out that the US government already stood against the use of cyberattacks against civilians in times of peace, which was the area we were trying to restrict. And more broadly, the history of weapons technology showed that even if the United States was in a leadership position today, other nations would catch up soon.

They pointed out that if we created stronger rules and the United States followed them, the country's adversaries would not. But we believed international rules could nonetheless help put greater pressure on all countries, including by creating the moral and intellectual foundation needed for more coordinated international responses to cyberattacks. After all, it was even harder to restrain conduct if it violated no rules in the first place.

As always, we learned a lot from these exchanges. Some people pointed out that important international norms were already in place and that we risked creating a perception that existing rules didn't matter. They were right. We were clear from the outset that we saw a Digital Geneva Convention as a long-term goal and part of a vision that would probably take up to a decade to implement. We didn't want to undermine existing norms along the way. We talked through these aspects in more detail with government and academic experts from around the world, acknowledging what rules already applied in cyberspace and the need to both strengthen their application and identify gaps that need to be filled.[8]

We also encountered pushback from people who objected to the notion that international companies would protect civilians on a global basis rather than help their home government attack other nations. As one Trump adviser challenged me on a trip to Washington, DC, "As an American company, why won't you agree to help the US government spy on people in other countries?"

I pointed out that Trump Hotels had just opened a new property in the Middle East as well as down the street on Pennsylvania Avenue. "Are these hotels going to spy on people from other countries who stay there? It doesn't seem like it would be good for the family business." He nodded.

At least we had succeeded in sparking a new conversation. When Satya and I attended a tech summit at the White House in June 2017, I participated in a breakout session on cybersecurity issues. One of the White House officials relayed a message to me beforehand: "Please don't raise the topic of a Digital Geneva Convention. We want this discussion to focus on security best practices for the US government rather than other issues."

As we walked into the ornate conference room where the meeting would take place, I reassured him that I got the message. But as the discussion ensued, the CEO of another company, whom I hadn't even spoken with, suddenly leaned across the table and said, "Look, what we really need is a Digital Geneva Convention."

I exchanged glances with the White House staffer and shrugged.

As we talked with more people about the notion of a Digital Geneva Convention, we realized that many of the points raised were pertinent to any form of arms control. There was a long history of public discussion about rules to govern weapons, and we needed to learn from them.

During the latter decades of the Cold War, arms control was *the* geopolitical focus as the world's superpowers of the time—the United States and the Soviet Union—negotiated treaties to manage nuclear weapons.[9] Issues around arms control were well understood in policy circles at the time, and they were often discussed much more broadly as well. As the possibility of a man-made nuclear apocalypse dwelled in the recesses of people's minds, it burst into pop culture in a big way in the early 1980s.

These nuclear risks weighed heavily on President Reagan on June 4, 1983, as he helicoptered to Camp David in rural Maryland with a stack of classified arms control documents. As a storm rolled into the Appalachians that evening, Reagan, with his wife, Nancy, settled into the lodge for a movie—one of the 363 films the former movie star would watch during his two-term presidency.[10] A writer for the new film, *WarGames*,[11] had arranged a screening; it had premiered the day before.

The thriller features a teenage hacker who goes from changing his grades in the high school's computer to stumbling into a super-computer at the North American Aerospace Defense Command, or NORAD, and nearly starting World War III. The Cold War tale spooked the commander in chief. Two days later during a high-level meeting at the White House, he asked whether anyone had seen the movie. Receiving blank looks, he described the plot in detail, before asking the chairman of his Joint Chiefs of Staff if the plot line was plausible.[12] That conversation set off a chain of decisions that led to the first federal forays into cybersecurity. Life imitated art, leading in part to the passage of the Computer Fraud and Abuse Act to make the hacking portrayed in the movie illegal.[13]

WarGames stoked the era's unease about nuclear weapons and technology. At a time when personal computers were nascent devices mostly sequestered to hobbyists' back bedrooms, the movie had wide appeal. Filmed more than thirty-five years ago, it now seems almost prophetic. Its themes connect with the public's concerns about computer vulnerabilities, the threat of war, and the prospect of machines escaping human control. They also speak to the power of diplomacy over war, captured by NORAD's supercomputer, which applied its learning from playing tic-tac-toe to the destruction that would be unleashed from nuclear war, uttering the film's climactic line: "A strange game. The only winning move is not to play."

Since the end of the Cold War, the topic of arms control has in many ways receded from public view. As a result, a generation of arms control experts has departed the scene and there is no longer a broad public understanding of the issues. In 2018, we were once again looking back to the future. As former US ambassador to Russia Michael McFaul put it, there was no longer a Cold War, but instead a Hot Peace.[14] It's time to dust off some of the lessons from the past.

In some respects, the response to World War II and to decades of nuclear arms negotiations provide some inspiration for the work needed to address cybersecurity. After all, following the dropping of two atomic bombs on Japan in 1945, the world has avoided nuclear conflict for almost seventy-five years. There are lessons from the challenging and at times circuitous paths taken by governments between the end of World War II and the end of the Cold War.

One such lesson came from International Humanitarian Law and the work of the world's governments when they came together in 1949 to create the Fourth Geneva Convention. The outcome didn't ban or limit specific weapons as much as it restrained *how* governments engage in military conflict. Under its rules, governments cannot deliberately target civilians, take action that would create disproportionate civilian casualties, or use weapons that cause superfluous injury beyond their military value.[15] Interestingly, the driving force for the 1949 Convention was not a specific government but rather the International Committee of the Red Cross, which continues to play a vital role in the Convention's implementation to this day.[16]

In an important way, the Geneva Convention speaks to learning that's applicable to arms control itself. It's often more realistic to limit the amount or characteristics of specific weapons or control how they're used than it is to try to ban them entirely. As one author suggested, "If a weapon is seen as horrific and marginally useful, then a ban is likely to succeed. If a weapon brings decisive advantages on the battlefield, then a ban is unlikely to work, no matter how terrible it may seem."[17]

Arms control ranks among the world's more difficult endeavors. But as one study concluded just as the Cold War was ending, agreements to control weapons so they are not used—as distinct from eliminating them entirely—"may, in the end, be better, if only because its prospects for success are greater."[18] It's perhaps this concept, as much as anything else, that has animated the efforts of international legal experts to define international norms that limit the way cyberweapons can be used.[19]

Another repeated lesson from the history of arms control is also applicable: Governments will sometimes seek to evade international agreements if they can, so there need to be effective ways to monitor compliance and hold violators accountable. This speaks directly to one of the biggest challenges for controlling cyberweapons. Governments not only perceive them as useful but also uniquely easy to use in a manner that evades detection. As David Sanger of the *New York Times* has put it, this unfortunately makes them "the perfect weapon."[20]

This points to the importance of increasing the ability to attribute cyberattacks to the nations that launch them and developing a collective capability to respond when such attacks take place. The United States and other governments are increasingly working to develop such responses, which can vary from responsive attacks to more conventional diplomatic tools, including sanctions. But regardless of the form, they are likely to contribute best to cyber-stability if they are grounded in agreements on what international rules are being violated and in multilateral consensus on who was responsible for the attack. And in an era when these new weapons are unleashed on data centers, cables, and devices owned and operated by companies, information from the private sector is likely to play a broader role in attributing attacks in the first place.[21]

All of this points to the continuing importance of international diplomacy. As we think about this new generation of diplomatic challenges, there are some new tools in the diplomatic toolbox. The Danish foreign minister identified one of our new opportunities when he said that tech companies have become a "nation" of sorts. While we thought the comparison had its limits, it underscored a key opportunity. If our companies are like nations, then we can forge our own international agreements.

We had tried to move the tech sector in this direction when we called for a "neutral digital Switzerland." We needed to put this into action by bringing companies together to sign an accord committing to act to defend all of our legitimate customers everywhere. While we sensed broad support for our general cybersecurity concepts, we knew this would not be an easy task. The tech sector is full of energetic people working for ambitious enterprises. Bringing companies together to do something in a coordinated way is easier said than done.

Establishing what would be called the Cybersecurity Tech Accord[22] was the perfect task for our digital diplomacy team, led by Kate O'Sullivan. She leads a team of Microsoft "diplomats" who work with policy makers and industry partners around the world to advance trust and security on the internet. Given the private ownership of cyberspace, we had long recognized that protecting it required not only multilateral but multi-stakeholder engagement. Like the new tech ambassadors representing governments, we needed envoys steeped in diplomatic values to focus on creating digital peace—and protecting our interests and customers in what had become a new plane of war.

We sketched out the tech accord's principles, and the digital diplomacy team fanned out to explore the industry's interest. The accord would first commit all its signatories to two overarching concepts: to protect users and customers everywhere and to oppose attacks on innocent citizens and enterprises from anywhere. These would provide the principled foundation that we thought the tech sector needed to promote and protect cybersecurity on a global basis. The agreement would complement these two principles with two pragmatic pledges. The first was to take new steps to strengthen the technology ecosystem by working with users, customers, and software developers to strengthen security protection in practical ways. And the second was to work more closely with each other to promote cybersecurity, including by sharing more information and coming to each other's aid when needed to respond to cyberattacks.

Getting people to agree that these principles made sense was one thing. Getting them to commit to them publicly was another. A small group came together quickly, including Facebook, which was becoming more forward leaning as it addressed its own rising privacy concerns. Several other large and experienced IT firms, including Cisco, Oracle, Symantec, and HP, also were quick to support the cause.

Google, Amazon, and Apple proved more difficult. As we talked with them, some said they considered it too controversial to stand next to Facebook at a time when it was on the firing line in national capitals around the world. Having experienced our own time on the firing line in the 1990s, I probably had more sympathy for Facebook than most. I also appreciated that to some degree, everyone has their difficult days, and if our first principle is to run away from people in times of trouble, we may doom ourselves to inaction even on issues for which collective efforts are needed the most.

Others at these companies said they had heard some of the push-back from individuals in the US government. They didn't want to support something that was the subject of criticism. And some said that they simply couldn't get people within their company to make a decision, so they couldn't get the approval to sign on. Despite repeated emails and phone calls, we couldn't get these companies across the finish line.

The good news was that the rest of the industry began to jump on board. We decided internally that we would launch the tech accord if we could secure public signatures from at least twenty companies. As the date for the RSA conference in 2018 in San Francisco approached, it was apparent we would meet that goal.

In the final weeks running up to announcing the Cybersecurity Tech Accord, we shared our plans with the White House and key officials in other parts of the United States and other governments. We didn't want them to be surprised. We got positive feedback from the White House itself, but we heard through the grapevine that some in the intelligence community had concerns about the language pledging not to help governments launch cyberattacks against "private citizens and enterprises." They were concerned that the reference to "private citizens" would cover terrorists and mean they could not turn to the tech sector if there was such an emergency. It was helpful feedback. We changed the text to refer to "innocent citizens" instead, which seemed to address this issue.

When we unveiled the Cybersecurity Tech Accord in April 2018, thirty-four companies signed on.[23] It was more than enough to generate momentum. By May 2019, the group had more than one hundred companies from more than twenty countries, and it was putting the accord into action by endorsing practical steps to strengthen cybersecurity protection.

Importantly, the need for stronger private sector collaboration found additional support around the world. To its credit, Siemens led one of the earliest efforts, creating what it called a Charter of Trust to focus on protecting the ubiquitous small devices that make up the internet of things. A number of leading European companies, including Airbus, Deutsche Telekom, Allianz, and Total, were quick to join.[24]

In some ways, an even more interesting reaction awaited us in Asia. While in Tokyo in July 2018, we met with senior executives at Hitachi, which wanted to be the first large Japanese signatory. When we arrived at their headquarters to seal their approval, they were quick to say, "We were attacked by WannaCry. We thought about staying silent, but we realized that we'll never solve this problem if we don't stand up together and do something like this."

This was in fact the whole point. I was struck by the fact that a long-standing Japanese technology company in a sector that had a reputation for being more conservative than American companies was willing to stand up at a time when companies like Google, Apple, and Amazon were still sitting down. We discussed in Tokyo the need for the tech sector to be proactive and build new forms of multilateral alliances.

Our preference was to see more government leadership that would sustain the multilateral approach to security that had been fundamental since the end of World War II. But it wasn't the mood of the moment in the White House or in other national capitals that were turning inward.

It seemed ironic and even uncomfortable as a company to advance the mantle of multilateralism, which typically was the role of governments. But we found far more support than criticism as we pushed forward. And as we made progress, an increasing number of companies expressed a desire to join in.

But if the diplomacy was to be effective, we'd need to take it beyond the tech sector and business community. Governments, companies, and nonprofit groups would need to find a way to act together. We looked for the right opportunity and concluded that our best chance was an international conference that would take place in Paris in November 2018. French president Emmanuel Macron had decided to host what he called the Paris Peace Forum on the centennial of the armistice that had ended World War I. He posted a video on YouTube that we watched several times.[25] It talked about the weakening of democracies and the collapse of multilateralism in the two decades that followed the armistice, leading to World War II. Macron invited ideas for projects that would strengthen democracy and multilateralism in the twenty-first century. It seemed like the perfect invitation for what we wanted to do.

Officials in Paris took an interest. David Martinon, France's ambassador for cyber diplomacy and the digital economy, held a position similar to that of Casper Klynge in Denmark. He was responsible for internet governance, cybersecurity, freedom of expression, and human rights. Under the leadership of Philippe Étienne, diplomatic Advisor to President Macron, Martinon and other French officials were already focused on architecting the future. We talked with them about the opportunity to craft a new declaration and initiative to address cybersecurity.

It took strong French leadership and months of careful conversations around the world. The day after the armistice centennial, President Macron unveiled the "Paris Call for Trust and Security in Cyberspace,"[26] reinforcing the importance of existing international norms to protect citizens and civilian infrastructure from systemic or indiscriminate cyberattacks. It also calls on governments, tech companies, and nongovernmental organizations to work together to protect democratic and electoral processes from nation-state cyberthreats—an area that we believed called for more explicit support under international law.

Even more important was the breadth of support for the Paris Call. The afternoon of Macron's speech, the French government announced that there were 370 signatories for it. The list included 51 governments from around the world, including all 28 members of the European Union and 27 of the 29 NATO members. It also included key governments from other parts of the world, including Japan, South Korea, Mexico, Colombia, and New Zealand. By early 2019, that number would rise to more than 500 and include 65 governments and most of the tech sector, including Google and Facebook—although not Amazon or Apple.[27]

Ironically and in our view unfortunately, the Paris Call garnered all this support without the backing of the United States government, which didn't sign the declaration in Paris. Although we originally were hopeful that Washington would sign on, it became apparent a month before the Paris meetings that the U.S. government wasn't ready to take a position one way or another. The political winds among some on the White House staff were not blowing in favor of multilateral initiatives, regardless of the issue. It put us in an unusual position, as we had our government affairs teams around the world asking other countries to support the effort.

The Paris Call nonetheless represents an important innovation. It takes the multilateral approach that had been so vital to international peace in the twentieth century and turns it into the type of multi-stakeholder approach needed to address global technology issues in the world today. It unites most of the world's democracies and connects them with most of the tech sector and leading nongovernmental groups worldwide. And over time, additional parties can sign on.

The model embodied in the Paris Call soon garnered additional global attention. As we met with Prime Minister Jacinda Ardern and her cabinet in New Zealand shortly after the Christchurch tragedy in March 2019, we brainstormed how the world could prevent a recurrence of terrorists using the internet as a stage for the attack on her people. Our conversation quickly turned to the Paris Call and whether we could bring governments, the tech sector, and civil society together in a similar way. We thought about it overnight, and by the next morning at a meeting with additional government officials, the room was talking about what a "Christchurch Call" might address.

Under Ardern's leadership, the New Zealand government seized the initiative. As I had commented to Ardern in our initial meeting, she brought to the issue a sense of moral authority. She was quick to reply that the world's outrage would eventually dissipate, and she wanted to use the moment not to score public relations points but to achieve something of more lasting importance. She dispatched New Zealand cybersecurity official Paul Ash to Europe to explore partnering with a government there, and building on the Paris Call, he found Macron's team willing to move briskly.

The tech industry had a big part to play. Our challenge was identifying pragmatic steps we could take to help prevent our services from being used as they were in Christchurch to amplify extremist violence. Inside Microsoft, I asked general counsel Dev Stahlkopf and her chief-of-staff, Frank Morrow, to spearhead work to develop ideas. Although we hadn't experienced the broad video uploading that had impacted Facebook, Twitter, and Google's YouTube service, we soon concluded that we had nine distinct services that potentially were susceptible to this type of abuse. These ranged from LinkedIn and Xbox Live to the sharing of videos through OneDrive, Bing search results, and the use of our Azure cloud platform.

Other tech companies were prepared not only to step forward but to step up. Google, Facebook, and Twitter all recognized that the use of their content sharing services by the Christchurch terrorist made it essential for them to do more. Amazon, to its credit, recognized that it could be part of the solution even if its services hadn't been part of the problem.

It was apparent that a variety of steps would be needed—and different balances would need to be struck—on different technology services. We needed to be sensitive both to engineering requirements and broader human rights and free expression concerns. A series of group conference calls quickly garnered support among the companies for nine concrete recommendations to address extremist violence and terrorist content online, including five steps individual services could take to tighten their terms of service, better manage live videos, respond to user reports of abuse, improve technology controls, and publish transparency reports. The group also formulated four industry-wide steps, including the launch of a crisis response protocol, open source–based technology development, better user education, and support for research and broader work by nongovernmental organizations to promote pluralism and respect online.

Ardern pressed for a decision and announcement at an upcoming meeting in Paris, only a month away. Representatives from the New Zealand and French governments met in northern California with civil society groups and with tech companies to talk through the specific issues raised by the proposed draft for the Christchurch Call. The New Zealand government's team worked pretty much around the clock, juggling feedback from government leaders and other stakeholders. On a late-night phone call that Satya and I had with

Ardern, I mentioned how struck I was by the government's speed. She replied, "When you're small, you have to be nimble!"

On May 15, two months to the day after the horrific attacks in New Zealand, Ardern joined Macron in Paris with eight other government leaders to launch the "Christchurch Call to Action." Its text addresses terrorist and violent extremist content online through commitments by governments and tech companies to act both separately and together.[28] As I joined other tech leaders and the heads of state in Paris to cement Microsoft's signature, our group of five companies also unveiled the nine steps we would take to put the Christchurch Call into action.

Launched just a little more than six months apart, the Paris and Christchurch calls highlight the progress the world can make by advancing what Casper Klynge likes to call "techplomacy." Instead of relying on governments alone, a new approach to multi-stakeholder diplomacy brings governments, civil society, and tech companies together.

In some ways, the idea is not entirely new. As one recent study concluded, a variety of nongovernmental organizations have long played influential roles on arms control issues, including involvement by advocacy groups, think tanks, social movements, and education groups.[29] Initially spearheaded in the 1860s by the founders of the Red Cross in Geneva, one of the most successful recent initiatives was the International Campaign to Ban Landmines in the 1990s. The latter campaign started with six nongovernmental organizations in 1992 and grew to involve roughly 1,000 such NGOs from sixty countries.[30] The group "successfully reframed landmines into a humanitarian and moral issue rather than a purely military matter" and, with support from the Canadian government, took its campaign to an ad hoc forum that adopted "a landmine-ban treaty in December 1997, barely five years after the campaign for a ban was initiated."[31]

Viewed from this perspective, the most novel aspect of the Paris and Christchurch calls is perhaps the involvement of companies, as distinct from other types of non-state actors, on a new generation of humanitarian and arms limitation issues. There is no doubt that some will be more skeptical of companies than of NGOs. But given the degree to which cyberspace is owned and operated by these companies, it seems hard to argue that they have no role to play.

The Paris and Christchurch calls also pointed to another innovation that we felt was important to ushering in an era of digital diplomacy. Arms control and humanitarian protection have always required broad public support. In the twentieth century, new ideas sometimes moved successfully from think tanks to detailed conversations with nongovernmental organizations and government policy circles, ultimately breaking through to the public by way of important speeches by international statesmen. But at a time dominated by the fragmentation of the traditional press and the rise of social media, there's both a need for and opportunity to connect with the public in new ways.

This is part of what we took away from the public discussion about a Digital Geneva Convention. While some traditional diplomats might roll their eyes, the idea captured public imagination in a way that had escaped the expert discussion of the critical but less than glamorous sounding international cybersecurity *Tallinn Manual 2.0.*[32] It's part of what I saw in the innovative approach and frequent tweets by Casper Klynge.[33] And it's what inspired us to couple our work on the Paris Call with support for citizen diplomacy, including an online pledge that garnered more than 100,000 signatures from around the world to support "Digital Peace Now."[34]

Perhaps as much as anything, we need to advance digital diplomacy with a sense of determination based not just on new circumstances and hopeful lessons from the past, but history's sobering failures as well. We were reminded of this when we visited the United Nations European headquarters in Geneva for a speech in November 2017. The Palais des Nations, which now houses the UN, served as the headquarters for the League of Nations in the 1930s. The building still has several small art deco conference rooms that reflect the post–World War I era.

The building served as the global stage for what would become some of the twentieth century's most tragic times. Japan invaded Manchuria in 1931, and shortly afterward, Hitler's Nazi regime became an increasing threat in Europe. Governments from thirty-one nations came together in the building to seek to constrain arms build-ups in a series of meetings that spanned more than five years. But the United States balked at providing the leadership needed for what it perceived were mostly European issues, and Hitler pulled Germany out of the negotiations and then the League of Nations itself, sounding the death knell for the effort toward global peace.

Before the diplomatic conference convened in 1932, Albert Einstein, the greatest scientist of his age, proffered a warning that fell on deaf ears. Technology advances, he cautioned, "could have made human life carefree and happy if the development of the organizing power of man had been able to keep step with his technical advances."[35] Instead, "the hardly bought achievements of the machine age in the hands of our generation are as dangerous as a razor in the hands of a three-year-old child." The conference in Geneva ended in failure, and before the end of the decade, that failure had translated into unimaginable global devastation.

Einstein's words speak to the crux of today's challenge. As technology continues to advance, can the world control the future it is creating? Too often, wars have resulted from humanity's failure to keep pace with innovation, doing too little too late to manage new technology. As emerging tech such as cyberweapons and artificial intelligence become more powerful, our generation will be put to this test yet again.

If we're to succeed where people failed almost a century ago, we'll need a practical approach to deterrence combined with new forms of digital diplomacy. As we joined Casper Klynge and his colleagues from more than twenty governments at a meeting in San Francisco in April 2019, it was heartening to see a new generation of cyberdiplomats working more closely together.

There's no denying that Denmark is a small country, with a population of 5.7 million that makes it smaller than Washington state. New Zealand's population is even smaller. But the Danish foreign minister had been right. In the twenty-first century, the best way to address global issues is to put in place a team that can work not only with other governments but also with all the stakeholders that are defining technology's future. It would be a mistake to underestimate a small country with a good idea and determined leadership. A new type of digital diplomacy has arrived.

Chapter 8 ⟫⟫ CONSUMER PRIVACY:
"The Guns Will Turn"

In December 2013, as tech leaders met at the White House to urge President Obama to reform government surveillance practices, the conversation at one point changed course. The president paused and offered a prediction. "I have a suspicion that the guns will turn," he said, suggesting that many of the companies represented at the table held more personal data than any government on the planet. The time would come, he said, when the demands that we were making on the government would be made on the tech sector itself.

In many ways, it was surprising that the guns hadn't turned already. In Europe, arguably, they had turned long before. The European Union had adopted a strong data privacy directive in 1995.[1] It established a solid baseline for privacy protection that exceeded anything found in the United States. The European Commission built on this directive by proposing an even stronger privacy regulation in 2012. It would take four years of deliberations, but the EU adopted its sweeping General Data Protection Regulation in April 2016.[2] As the United Kingdom voted to leave the EU following its Brexit vote just two months later, data protection authorities in the country were quick to affirm their support for continuing to apply the new rules in the UK itself. As Prime Minister Theresa May told tech executives in a meeting in early 2017, the government recognized that the UK economy would remain dependent on data flows with the continent, and this required a uniform set of data privacy rules.

But what about the United States? As data privacy rules continued to sweep around the world, it remained an outlier. Across Europe officials fretted increasingly about the protection of their citizens' privacy in a world where data crossed borders and entered American data centers, but the United States failed to protect privacy broadly at the national level. I had given a speech myself on Capitol Hill in 2005 calling for the adoption of national privacy legislation.[3] But aside from HP and a few other companies, most of the industry either yawned or opposed the idea. And Congress remained uninterested.

It would take the efforts of two unlikely individuals to spark change in the United States. The first person to go to the mats on privacy was a law student from the University of Vienna named Max Schrems. During a stopover in Europe in 2019, Max introduced us to the Austrian delicacy of boiled beef and shared his unlikely story.

Schrems is a bit of a celebrity in Austria, and if you followed the trans-Atlantic privacy saga, he's instantly recognizable. "I lost my privacy over a privacy case," he laughed.

Privacy, including the American notion of it, has always intrigued him. When Schrems was seventeen, he was dropped into the "middle of nowhere" Florida as a high school exchange student. The tiny town of Sebring was a culture shock for sure, but not for reasons that you might expect. It wasn't social gatherings centered around the Future Farmers of America or the southern Baptist church that disoriented Schrems, but rather the school's methods of tracking students.

"There was a whole pyramid of control," he said. "We had a police station at the school and there were cameras in every corridor. Everything was tracked from grades, SAT scores, attendance, and little stickers on our student IDs that allowed us to use the internet."

Schrems proudly recalled how he helped his American peers circumvent the blocks the school had put on Google searches. "I showed them that there's Google.it, which works perfectly fine because the school only blocked dot-com," he said. "The exchange student introduced the school to international top-level domains!"

It was a relief to return to Vienna, he told us, "where we have so much freedom."

Privacy was still on Schrems's mind in 2011, when at the age of twenty-four he returned to the United States for a semester at the Santa Clara University School of Law in California. A visiting lecturer, who also happened to be a lawyer at Facebook, spoke to Schrems's privacy class. When Schrems quizzed him on the company's obligations under European privacy law, the lawyer replied that the laws weren't enforced. "He told us that 'you can do whatever you want to do' because the penalties in Europe are so trivial that the enforcement is nonexistent," Schrems said. "Obviously, he didn't know a European was in the room."

The exchange inspired Schrems to dig deeper and write his term paper about what he regarded as Facebook's shortcomings in addressing its European legal obligations.

For most students, the story would have ended there. But Schrems was no ordinary student. Within a year, he took what he learned and filed complaints with the data protection authority in Ireland, where Facebook's European data center was located. His complaint was straightforward but had implications that were potentially upending for the global economy. He argued that the International Safe Harbor Privacy Principles put in place to allow European data to be transferred to the United States needed to be struck down. The reason, he argued, was that the United States had insufficient legal safeguards to protect European data properly.

The Safe Harbor principles were a fundamental pillar of the trans-Atlantic economy, but it was little known except by privacy experts. It was a creature of the EU's 1995 privacy directive, which permitted Europeans' personal information to move to other countries only if they had adequate privacy protection in place. Given the absence of a national privacy law in the United States, some political creativity was needed to keep data moving across the Atlantic. The solution, adopted in 2000, was a voluntary program that enabled companies to self-certify that they complied with seven privacy principles endorsed by the US Department of Commerce. The principles mirrored EU rules, and they enabled the European Commission to conclude that the United States provided adequate privacy protection as required by the 1995 directive.[4] The International Safe Harbor Privacy Principles were born.

Fifteen years later, the trans-Atlantic movement of data had exploded. More than four thousand companies were taking advantage of Safe Harbor to deliver $240 billion of digital services annually.[5] This included everything from insurance and financial services to books, music, and movies. But the financial aspects represented just the tip of the information iceberg. American companies had 3.8 million employees in Europe who depended on Safe Harbor data transfers for everything from their paychecks to health benefits and personnel reviews.[6] The total European sales of American companies topped a whopping $2.9 trillion, and most of it required the movement of digital data to make sure that goods reached their destination and revenue was accounted accurately.[7] It was a barometer of the world's extraordinary reliance on data.

While government officials and business leaders saw Safe Harbor as a modern-day necessity, Max Schrems saw something entirely different. Like the youth in Hans Christian Andersen's fairy tale, he looked at the Safe Harbor principles and asserted, in effect, that "this emperor has no clothes."

Schrems had been a Facebook user since 2008 and relied on that basis for his complaint with the Irish Data Protection Commissioner. By 2012, he had returned to Vienna. After a "twenty-two-email back-and-forth" with Facebook, Schrems received a CD containing a PDF with twelve hundred pages of his personal data. "It was only half or a third of what they had on me, but three hundred pages of it were things I had deleted—it actually said 'deleted' on each post."

As he saw things, a Safe Harbor agreement that permitted Facebook to collect and use so much data in this way couldn't possibly provide the protection that European law required.

Schrems publicized his complaints, creating a minor media cycle across Europe as he argued that the Safe Harbor principles should be struck down. Facebook quickly dispatched two of its senior European executives to Vienna to try to persuade him to reconsider his views. They spent six hours in a hotel conference room next to the airport urging Schrems to narrow his complaints. But he would not drop them, insisting that he wanted the Irish commissioner to address his concerns.[8]

Others in the tech sector and privacy community followed the issue with interest, but most didn't expect Schrems's case to go far. After all, he had spent more time drafting his complaints than completing his term paper at Santa Clara, which remained unfinished but with an extension from his professor.[9] Soon the matter appeared to reach the end of the road, as the Irish Data Protection Commissioner ruled against Schrems, concluding that it was bound by the European

Commission's finding in 2000 that Safe Harbor was adequate. It seemed time for Schrems to return to writing his law school paper. But he would not back down.

His case finally reached the European Court of Justice. And on October 6, 2015, all hell broke loose.

I was in Florida getting ready for an event with customers from Latin America when the phone rang early in the morning. The court had struck down the International Safe Harbor Privacy Principles.[10] It concluded that Europe's national data protection authorities are empowered to make their own assessments of data transfers under the agreement. In effect, the court gave more authority to independent regulators it knew would be tougher in reviewing privacy practices in the United States.

Immediately people wondered whether this meant a return to the digital dark ages. Would trans-Atlantic data flows now stop? Preparing for precisely this contingency, we had put in place other legal measures to ensure our customers could continue to use our services to move their data internationally. We scrambled to reassure customers. Across the tech sector everyone put on a brave face, but the European court's decision was disturbing. In the words of one lawyer who helped negotiate Safe Harbor, "We can't assume that anything is now safe. The ruling is so sweepingly broad that any mechanism used to transfer data from Europe could be under threat."[11]

The decision led to months of feverish negotiations. It was a bit like trying to put Humpty Dumpty back together again. U.S. Commerce Secretary Penny Pritzker and European Commissioner Věra Jourová were on point to develop an approach that was more likely to satisfy the court and Europe's various privacy regulators. As I arrived at the European Commission in January 2016 to discuss the state of play with Jourová, I was surprised when she greeted me while I was downstairs waiting for an entrance pass. She chuckled as she mentioned that she had just stepped out for a moment. A gentleman she had never met recognized her outside and walked up and said, "We should know each other. My name is Max Schrems."

While the international negotiations continued, the tech sector prepared for the worst. At Microsoft we explored whether we could take advantage of Seattle's proximity to Canada and shift key support to our facilities in Vancouver. It would mean shuttling some Redmond employees back and forth, but because the court's decision didn't impact data transfers between Canada and Europe, we could ensure more seamless operations.

Ultimately, this proved unnecessary. In early February 2016, Pritzker and Jourová unveiled a new agreement. They replaced the Safe Harbor principles with the Privacy Shield, which contains heightened privacy requirements and an annual bilateral review. Microsoft became the first tech company to pledge that it would comply with the new data-protection demands.[12]

A data disaster had been averted. But the episode revealed just how much had changed.

For one thing, it showed that there was no such thing as a privacy island—no longer could anyone assume that all their data stayed within the borders of a single nation. This is the case even for a continent as big as Europe or an economy as large as the United States. Personal information moves from country to country for all types of digital transactions, and most of the time people are not aware of it.

This created a new outside political lever with potentially profound privacy implications for the United States. The members of the European Court of Justice effectively empowered the continent's data-protection regulators, well known for their passionate commitment to privacy, to negotiate for tougher American privacy standards.

If there was any doubt about this goal, it was dispelled when credible firsthand reports circulated quietly within government circles shortly after the court's 2015 decision. A member of the court who had been at the center of its deliberations met in person with several national privacy regulators to walk them through the details of the decision and recommend how they could best use it to negotiate with the White House and Department of Commerce. It was the type of step that would fly in the face of the separation between courts and the executive branch in the United States. It was unusual in Europe, but not unheard of in many parts of the world.

While American political leaders can give speeches that denounce the overreach of European privacy regulators, there is one thing they cannot change. This is the critical reliance of the US economy on the ability of American firms to move data to and from other countries. In the world today, one can debate whether to construct an immigration wall to stem the flow of people. But no nation can tolerate a barrier that stops the international flow of data. This means trans-Atlantic negotiations that impact the privacy practices of American companies have become an economic fact of life.

Even the ultimate implications for China are weighty. Over time, the European approach can lead to mounting pressure on China to confront an important crossroads. It can move forward without privacy protection for data within its borders, or it can strengthen its economic connections with Europe with the inevitable data flows this will require. But it will become more difficult to do both.

Like the reaction to many near-disasters, however, the immediate response to the negotiation of the Privacy Shield was mostly a sigh of relief. It was another wake-up call, but again people hit the snooze button. Data flows could continue, and companies could continue to do business. Most tech companies and government officials postponed deeper thinking about the longer-term geopolitical implications for another day.

In some key respects, this was more than understandable. The rest of 2016, with the Brexit vote and the American presidential election, commanded people's attention. And within a few months, everyone was focused on a different privacy development from Europe. It was the EU's looming implementation date for its General Data Protection Regulation, or GDPR.

The GDPR quickly became a household acronym for people working in the tech sector. While it wasn't unusual to hear lawyers use acronyms to refer to government regulations, the GDPR started to roll off the tongues of engineers, marketers, and salespeople alike. There was good reason. The regulation required the re-architecting of many of the world's technology platforms, and this was no small task. While it wasn't necessarily part of the EU's plan, it became a second way for Europe to influence privacy standards across the United States and around the world.

The GDPR is different from many government regulations. Most of the time, a regulation tells a company what it cannot do. For example, don't include misleading statements in your advertisements. Or don't put asbestos in your buildings. The fundamental philosophy of a free market economy encourages business innovation, with regulation putting certain conduct off-limits but otherwise leaving companies broad freedom to experiment.

One of the biggest features in the GDPR is in effect a privacy bill of rights. By giving consumers certain rights, it requires that companies not just avoid certain practices but create new business processes. For example, companies with personal information are required to enable consumers to access it. Customers have a right to know what information a company has about them. They have a right to change the information if it's inaccurate. They have a right to delete it under a variety of circumstances. And they have a right to move their information to another provider if they prefer.

In important ways, the GDPR is akin to a Magna Carta for data. It represents a critical second wave of European privacy protection. The first wave came in 1995, with a privacy directive that required that websites notify consumers and get their consent before collecting and using their data. But as the internet exploded, people were inundated with privacy notices and had little time to read them. Recognizing this, Europe's GDPR required that companies give consumers the practical ability to go online to view and control all the data that had been collected from them.

It's not surprising that its implications for technology are so sweeping. Start with the proposition that any company with millions of customers—or even thousands of customers—needs a defined business process to manage these new customer rights. Otherwise it will be swamped with inefficient and almost certainly incomplete work by employees to track down a customer's data. But more than that, the process needs to be automated. To comply quickly and inexpensively with the GDPR, companies need to access a customer's data in a unified way across a variety of data silos. And this requires changes to technology.

For a diversified tech company like Microsoft, the impact of the GDPR could hardly have been more intense. We had more than two hundred products and services, and many of our engineering teams had been empowered to create and manage their own back-end data infrastructures. There were certain similarities, but there were also important differences in the information architecture used in different parts of the company.

Quickly we realized these differences would be problematic under the GDPR. Consumers in the EU would expect a single process that would pull all their information across all our services so they could see it in a singular and unified way. The only way for this to happen efficiently would be for us to create a new and singular information architecture that would span all our services from one end to the other. In other words, from services like Office 365 to Outlook to Xbox Live, Bing, Azure, Dynamics, and everything in between.

In early 2016, we assembled a team with some of our best software architects. They had two years before the GDPR would take effect on May 25, 2018, but they had no time to spare.

The architects needed first to turn to lawyers, who defined what the GDPR required. With the lawyers, they then created a specification that listed all the technology features our services would need to enable. The architects then crafted a new blueprint for the processing and storage of information that would apply to all our services and make these features effective.

By the last week of August, the plan was ready for review at a meeting with Satya and the company's senior leadership team. Everyone knew that the blueprint called for a massive amount of engineering work. We'd need to move more than three hundred engineers to work full-time on the project for at least eighteen months. And in the final six months before the GDPR implementation date, the number would swell into the thousands. It represented a financial commitment in the hundreds of millions of dollars. It was not a meeting that anyone wanted to miss. Some cut short their vacation to be there.

The engineering and legal teams walked through the blueprint, timelines, and resource allocations. It impressed everyone. And in some ways, it surprised everyone. As the meeting progressed, Satya suddenly exclaimed with a bit of a chuckle, "Isn't this great?" He continued, "For years it has been next to impossible to get all the engineers across the company to agree on a single privacy architecture. Now the regulators and lawyers have told us what to do. The job of creating a single architecture just got a whole lot easier."

It was an interesting observation. Engineering is a creative process and engineers are creative people. When two software engineering teams approached the same problem in different ways, it could be enormously difficult to persuade them to reconcile their differences and develop a common approach. Even if the differences didn't go to the heart of a feature of fundamental importance, people tended to hold on to what they had created.

Given the large, diversified, and empowered engineering structure at Microsoft, this challenge was sometimes greater than at other tech companies. It had led us in the past sometimes to maintain for years two or more overlapping services, an approach that almost never turned out well. Apple, in contrast, had sometimes relied on its narrower product focus and Steve Jobs's centralized decision making to solve this problem. It was perhaps ironic, but European regulators were doing us something of a favor by defining a singular approach that required engineering compromise all around.

Satya signed off on the plan. Then he turned to everyone and added a new requirement. "As long as we're going to spend all the time and money to make these changes, I want to do this for more than ourselves," he said. "I want every new feature that's available for our use as a first party to be available for our customers to use as a third party."

In other words, create technology that could be used by every customer to comply with the GDPR. Especially in a data-dominant world, it made complete sense. But it also added more work. All the engineers in the room gulped. They left the meeting knowing they'd need to put even more people on the project.

The enormous technical requirements help explain a second dynamic that quickly emerged, one with important geopolitical implications. Once the engineering work was on track to comply with the GDPR, it was hard to be enthusiastic about creating a different technical architecture for other places. The costs and engineering complexity of maintaining differing systems are just too great.

This led to an interesting conversation with Canadian prime

minister Justin Trudeau in early 2018. As Satya and I met with him and some of his top advisers, we touched upon privacy issues, which remain an important topic with the Canadian public. As Trudeau mentioned the potential for changes in Canadian privacy law, Satya encouraged him simply to adopt the provisions in the GDPR. While this suggestion was met with some surprise, Satya explained that unless there was some difference that was of fundamental importance, the costs of maintaining a different process or architecture for a single nation seemed likely to outweigh the potential benefits.

Our greater enthusiasm for the GDPR put us in a different camp from some others in the tech sector, who sometimes tended to focus more on the parts of the regulation they found onerous. While there were parts of the GDPR that we found confusing or worse, we believed that one key to long-term success for the tech sector was sustaining public trust on privacy issues. This approach was another area born from our antitrust experience in the 1990s and the high reputational price we had paid. A more balanced approach on potentially contentious regulatory issues might strike some of our tech sector peers or even our own engineers as excessively diplomatic. But I felt that there were many days and issues that made it the wiser course.

Others in the tech sector nonetheless often pointed to the American public's ambivalence toward privacy as reason to ignore US regulatory pressures. "Privacy is dead," they'd say. "People just need to get over it."

I believed the privacy issue would be quiet until the day it was not. A firestorm could break out with little of the political foundation in place for a more thoughtful approach. The public ambivalence toward privacy reminded me of the nuclear power industry's experience decades earlier.

Throughout the 1970s the nuclear power industry had failed to engage in an effective public discussion about the risks associated with its technological advances, leaving the public and politicians alike unprepared for the meltdown that occurred at the Three Mile Island Nuclear Generating Station in Pennsylvania in 1979. As a result of the calamity, and unlike in other countries, the political fallout from Three Mile Island stopped American nuclear power construction in its tracks. It would take thirty-four years before construction would start on another nuclear power plant in the United States.[13]

I felt this was a historical lesson to learn from rather than repeat.

In March 2018, the privacy equivalent of Three Mile Island arrived when the Cambridge Analytica controversy exploded. Facebook users learned that their personal data had been harvested by the political consulting firm to build a database targeting US voters with advertisements designed to support Donald Trump's presidential campaign. While the usage itself violated Facebook's policies, the company's compliance systems had failed to detect the problem. It was the type of issue that draws plenty of criticism but leaves a company with no real defense. All it can do is apologize, which is what Mark Zuckerberg did.[14]

Within weeks the public mood shifted in Washington, DC. Instead of dismissing regulation, politicians and tech leaders finally were talking about it as an inevitability. But they failed to describe what they thought this regulation should do.

That answer would come from the other side of the country, near Silicon Valley itself. And this drama involved a second character who was as unlikely to play a leading role as Max Schrems had been.

It was an American named Alastair Mactaggart. In 2015, the San Francisco Bay Area real estate developer hosted a dinner party at his home in Piedmont, California, a leafy suburb across the bay from the Silicon Valley empires quietly dealing in private information. As Mactaggart quizzed one of his guests about his job at Google, he wasn't just dissatisfied with the answers, he found them terrifying.

What private data were tech companies collecting? What were they doing with it? And how can I opt out? If people knew what Google knew, the engineer replied, "They'd freak out."

That conversation over cocktails set the wheels in motion on a two-year, more than $3 million crusade. "It felt very important. I thought, 'Someone has to do something about this,'" Mactaggart told us almost three years later when we met in San Francisco. "I figured that the someone to do something might as well be me."

The father of three wasn't looking to take a swing at the tech industry. He was a successful businessman and a firm believer in free and open markets. After all, he'd made his money from soaring real estate prices in the tech-fueled region. But he was determined to make a difference, hoping to one day tell his kids that he'd helped protect something precious: our personal data.

In the age of what Mactaggart and some others call "commercial surveillance," our online searches, communications, digital location, purchases, and social media activity tell more about us than we probably want to share.[15] He concluded that this bestowed incredible power on a handful of companies. "You must accept their privacy terms, or you can't use their services," he said, referring to the free online tools we unwittingly pay for with our information. "But these are services that we rely on to live in the modern world. There really is no opting out."

This lack of oversight propelled him to recruit like-minded supporters and draft a new privacy law for California. "I live in a highly regulated world," he said, referring to the accepted regulation and building codes that govern real estate. "It's healthy. The law needs to catch up with tech or people will just continue to push the boundaries."

Mactaggart had learned plenty about the workings of government from his real estate experience. He was politically astute, recognizing that Silicon Valley opposition would likely make it as difficult to pass a law in Sacramento, the state's capital, as it would be to pass a federal law in Washington, DC. But California, like some other western US states, had a political alternative. These states, established in the middle and late 1800s, had constitutionally mandated processes that, with enough signatures, could put an initiative on the ballot for the voters to decide.

California's initiative process had changed the course of American history in the past. Four decades earlier, in 1978, the state's voters adopted Proposition 13 to limit taxes. The measure reduced property taxes in the state, but its broader impact was far greater. It helped fuel a public movement across the country that added momentum to Ronald Reagan's Presidential election in 1980 and stronger national pressure to reduce the size of government and cut taxes. It created a watershed political moment, reflecting in part the fact that one in every eight Americans lives in California.

If Cambridge Analytica could become the equivalent of Three Mile Island, could Alastair Mactaggart create the privacy equivalent of Proposition 13?

It quickly seemed likely that the answer was yes. Mactaggart gathered more than double the signatures needed to put the measure on the ballot. His polling said that 80 percent of voters started out supportive of his proposal. He was disappointed that 20 percent of voters were not, until his pollsters explained that they had never seen such high numbers. While well-funded initiative campaigns almost always lead to a closer outcome in the end, it was apparent that if Mactaggart was willing to spend more of his real estate millions on an effective campaign, he would have more than a good chance of success at the ballot box in November.

At Microsoft, we considered Mactaggart's initiative with mixed feelings. On the one hand, we had long supported privacy legislation in the United States, including at the federal level. Led by Julie Brill, a former FTC commissioner who now leads the company's privacy and regulatory affairs work, we had decided to take a different approach from the rest of the tech sector when the GDPR took effect in May 2018. Rather than make the regulation's consumer rights available only to people in the EU, we extended this to all our customers everywhere in the world. It made for some surprising insights. We quickly learned that American consumers were even more interested in putting these rights to work than Europeans, validating our sense that the arc of American history would ultimately bend toward the adoption of privacy rights in the United States.[16]

But we found the text of Mactaggart's draft initiative complicated and in some respects confusing. We worried that in some places it would lead to technical requirements that would differ from the GDPR for little good reason. These were the types of problems that could be remedied by a legislature and its detailed drafting process, but not by a take-it-or-leave-it proposition at the ballot box. The question was how to persuade everyone to move the effort from the November ballot to the statehouse without killing it along the way.

Other tech companies embarked on a fund-raising campaign to oppose the initiative. Silicon Valley recognized that success would likely require raising more than $50 million. We donated $150,000. It was enough to stay connected with the rest of the industry but not the type of money that would give the opposition effort too much momentum.

Ultimately, the large amount of funding needed for a California initiative campaign created an incentive for both sides to negotiate. Mactaggart was willing to sit down with key elected officials to help hammer out the legislative details. Some of the other tech companies had a hard time deciding what they wanted to do. But we dispatched two of our privacy experts to Sacramento, where they worked pretty much around the clock, going through the details with the legislative leaders and Mactaggart's team.

At the last possible minute, the legislature adopted the California Consumer Privacy Act of 2018, and Governor Jerry Brown quickly signed the measure. It was the strongest privacy law in the history of the United States. Like the GDPR, it gives the Golden State's residents the right to know what data companies are collecting on them, to say no to its sale, and to hold firms accountable if they don't protect personal data.

The national impact was felt almost immediately. Within a matter of weeks, even opponents who had long resisted comprehensive privacy legislation in Washington, DC, began to discover something akin to new religion. With the California floodgates open, it was apparent that other states would likely follow. Rather than face a patchwork of state rules, business groups started lobbying Congress to adopt a national privacy law that would preempt—or in effect overrule—California's law and other state measures. While much remained ahead, Mactaggart had successfully changed the country's consideration of privacy issues. It was a momentous achievement.

When we sat down with Mactaggart in San Francisco, it was impossible not to be impressed. It would have been easy to see him as a threat—an activist looking to rein in an industry that had become too powerful. Instead, we found a likable pragmatist who was thinking broadly about the future.

"This isn't over," he said. "We'll be talking about technology and privacy for the next hundred years. Just like we do with antitrust law more than a century after the Standard Oil case."

As a company that had survived antitrust turbulence eighty years after the DOJ had broken up Standard Oil, we easily understood the comparison. And ultimately, Mactaggart's historical comparison provides some of the most important food for thought.

The combination of the efforts of Max Schrems and Alastair Mactaggart reveals several important lessons for the future.

First, it's hard to believe privacy will ever die the quiet death that some in the tech sector predicted a decade or two ago. People have awakened to the fact that virtually every aspect of their lives leaves behind some type of digital footprint. Privacy needs to be protected, and stronger privacy laws have become indispensable. The day will come when the United States joins the European Union and other countries in applying a law like the GDPR.

We're also likely to see a third wave of privacy protection emerge over the next few years, especially in Europe. Just as the GDPR responded to the plethora of privacy notices that people didn't have time to read, we're already hearing concerns that people lack the time to review all the data that the GDPR is making available online. This is likely to prompt a new wave of governmental rules to regulate how data can be collected and used.

This also means the tech sector will need to apply more of its technical creativity to innovations that protect privacy while also enabling data to be put to good use. Some new technical approaches have already emerged, like the ability to pursue AI advances with data that remains encrypted and hence better at protecting privacy. But this is just a start.

Finally, the experiences of Schrems and Mactaggart speak to important strengths and opportunities for the world's democracies. The leaders of an autocratic government might look with alarm at the unpredictable ability of a law student and a real estate developer to upend the rules that govern some of the most powerful technologies of our time. But there's another perspective—and on balance it seems the better view. Schrems and Mactaggart used established judicial and initiative processes to redress what they regarded as a wrong. Their success speaks to the ability of a democratic society, when it works well, to adapt to a people's changing needs and move a nation's law where it needs to go with less rather than more disruption.

The integrated nature of the global economy and the long reach of Europe's privacy rules will create pressure even on countries like China to adopt strong privacy measures. In other words, Europe is not just the birthplace of democracy and the cradle of privacy protection. It's quite possibly the world's best hope for privacy's future.

Chapter 9

⨠ RURAL BROADBAND:
The Electricity of the
Twenty-first Century

When you approach the Knotty Pine Restaurant & Lounge on the main drag in Republic, Washington, you know you're in a former Old West boomtown. The cedar false front gestures to an age when ramshackle mining and logging towns were built in a hurry. But the bright yellow placard hanging at the door declaring "Bikers Welcome" reminds you that this small town is squarely in a new era.

We'd spent the morning taking a left instead of a right, and then a right instead of a left, driving our rental car on a circuitous path down ranch and farm roads. It's a beautiful place, so we didn't mind the detour. But we had an appointment to keep and our GPS had proven useless in the northeastern corner of the state. We finally ditched our smartphones and followed State Route 20 on a paper map to Ferry County and straight into town, where a small group of locals were expecting us before lunch.

Ferry County persistently has the highest unemployment rate in the state, topping 16 percent when the region's agricultural work falls with the winter temperatures. We'd come from King County, home of Microsoft, Amazon, Starbucks, Costco, and Boeing. With an unemployment rate that hovers well below 4 percent, King County has led the state in growth that boasts twice the national average. We wondered how Ferry County could tap into the twenty-first-century boom taking place on the other side of the Cascade Mountains.

The Knotty Pine promised breakfast all day, so for $5.95 we ate scrambled eggs, bacon, and maple-syrup-drenched pancakes so big they draped onto the table. After our long drive, the food was welcome, but it was quickly dimmed by the hospitality.

Originally called Eureka Gulch by the prospectors who founded the small town at the end of the nineteenth century, Republic rests in the valley between the pine-draped Wauconda and Sherman passes of northeastern Washington. The Sherman Pass is named after the famous Civil War general William Tecumseh Sherman, who traveled across it in 1883. It's breathtaking country and an outdoor enthusiast's dream.

The town's past was defined by the golden veins that snaked through the surrounding granite and bore deep into the river gulches. Banking, transportation, and other supporting services soon followed the miners—and later the loggers—to town. Today, the mines are closed, and the region is struggling to redefine its future.

When we arranged our meeting, Elbert Koontz, a former logger and now the town's mayor, told us that he would put on his "best sweats" for our lunch. I was disappointed when he instead showed up in pressed pants. While Elbert was generous with his knowledge and quips, he didn't laugh when I asked about the state of high-speed broadband in Ferry County. He just rolled his eyes.

"Almost no one around here has broadband," he said. "Promises have been made for years, but nothing's been done." But according to data from the Federal Communications Commission, or FCC, every person in Ferry County had access to broadband.

Thanks to a fiber-optic line traversing the Sherman Pass, the tiny town's center, home to around a thousand people, has some broadband access. But it was clear that the rest of the region's residents were lacking. "The problem is that here, we live up in the woods," Elbert told us. "I mean, when you go out of town, you're out of town, folks."

The mood shifted as the crowd nodded in agreement and shared their stories of broadband woes. Some relied on spotty satellite connections. Others came to town to tap into hotspots even just to download software upgrades for their laptops. Others were hopeful 5G would save the day. But it was unanimous around the table: The vast majority of Ferry County didn't have reliable high-speed broadband access.

"Tell *that* to the FCC," someone scoffed.

In fact, that's what we did.

A few months later, we dodged rain and traffic on Twelfth Street as we made our way to the FCC headquarters in Washington, DC. We had come to meet with FCC chairman Ajit Pai. After we checked in and passed through the security screening, we were ushered into his office.

Chairman Pai greeted us with a smile. "Thanks for coming! What can I do for you?"

Family photos lined his shelves and the windowsills of his office overlooking the soggy capital. He's a first-generation Indian American who was raised in Kansas by doctors who immigrated to the country just two years before he was born.

I told him about our trip to Ferry County and the situation on the ground. The FCC's national map shows that everyone in Ferry County has access to broadband. Everyone.

To Pai's credit, the FCC is focused on bringing broadband to every American, but it's a daunting and expensive problem to solve, especially when you don't truly understand how big the issue is. "You're not the FCC chairman who created this problem," I said, referring to the flawed data. "But you can be the chairman who fixes it." It needs to be treated as a national priority.

Like the Mayor of Republic had described to us, much of the federal government's data about Ferry County—and rural America, for that matter—is wrong. People throughout Ferry County were aware of this, which did little to instill confidence in their government. For these people, the inaccurate data is more than a small inconvenience. It impacts the allocation of federal funding for broadband, which doesn't flow to areas that the government believes have access already. And even more broadly, this lack of access impairs other critical public resources, such as those needed to fight wildfires that can rage through the West each summer.

"It's the Wild West out here," Elbert said. "We don't have a huge sheriff's department, we don't have a huge fire department, we don't have any of that. All our firefighters are volunteers." And those volunteers find themselves in perilous conditions when fires ravage the landscape.

In 2016, a fast-moving fire broke out when hot August winds ripped down a power line and fanned the resulting flames in northern Ferry County. Within five hours, an inferno had consumed more than twenty-five hundred acres and was growing.[1] Portions of the affected community were under a Level 3 evacuation, which translates to "get out now."

The spotty cellular infrastructure and lack of broadband made it impossible to transmit critical data to and from the fire lines to keep authorities updated on where the fire was headed and who needed to evacuate. The only way to share crucial information between firefighters, the forest service, and law enforcement was to load data onto a memory stick, hand it to a driver in a pickup truck, and wait for the truck to travel forty minutes from the fire line into Republic, where authorities could use the broadband and radio connections.

When a fire can turn a 20-mile-an-hour wind into a 50-mile-an-hour gale in a minute, according to Elbert, "that's just plain dangerous."

The Americans stranded in the dial-up era aren't confined to Ferry County. They are in every single state in the country. According to the FCC's 2018 broadband report, more than twenty-four million Americans, more than nineteen million of whom live in rural communities, lacked access to fixed high-speed broadband.[2] That's roughly the population of New York state.

This lack of broadband in rural communities isn't a question of affordability—these people can't buy the service if they want it. Many rely on dial-up technology to transmit data over copper lines, unable to access online services most of us take for granted at basic download and upload speeds.[3] In other words, a significant portion of rural communities lack the internet speeds that were available in urban areas over a decade ago.[4]

While that figure is sobering, there is strong evidence that the percentage of Americans without broadband access is much higher than the FCC's numbers indicate. As we analyzed this data, we found that it was based on flawed methodology. The FCC concludes that a person has access to broadband if a local service provider reports that it could provide such service "without an extraordinary commitment of resources."[5] But many such companies don't provide this service in practice. It's like telling someone they have access to a free lunch if a local restaurant says it could serve it to them if it wanted. It doesn't mean the restaurant will.[6]

In fact, other data paints a very different picture of the country. For example, the Pew Research Center has been tracking internet usage since 2000 through regular surveys. According to its latest data, 35 percent of Americans report that they don't use broadband at home—or roughly 113 million people.[7] And even the FCC's own subscription data indicates that 46 percent of American households fail to subscribe to the internet at broadband speeds.[8]

While there's a difference between broadband availability and usage, this difference is so large that one must ask whether one of these figures is just plain wrong. We asked our own data science team to do more detailed work based on public and Microsoft data sources. Their research suggests that the Pew numbers are much closer to the mark than the estimate from the FCC.[9] But even more than that, this leaves us with the inescapable conclusion that today there exists no fully accurate estimate of broadband availability in the United States, anywhere.

Does this matter? You bet, and in a big way.

Broadband has become the electricity of the twenty-first century. It's fundamental to the way people work, live, and learn. The future of medicine is telemedicine. The future of education is online education. And the future of agriculture is precision farming. Even with a future where there is more computing intelligence "at the edge"—meaning with ubiquitous small and powerful devices that process more data themselves—there still needs to be some high-speed access to the cloud. And that requires broadband.

Today, rural areas that lack broadband are still living in the twentieth century. And it shows in almost every economic indicator. Our data science team confirmed what universities and research institutions around the world have been finding: The highest unemployment rates in the country are frequently located in the counties with the lowest availability of broadband, highlighting the strong link between broadband availability and economic growth.[10]

When you talk to business leaders about where they might

expand their operations and add jobs, this requirement surfaces almost immediately. Asking them to open a new facility in a place without broadband is like asking them to set up shop in the center of the Mojave Desert. In a world reliant on modern high-speed access to data, an area without broadband is a communications desert.

The lack of job growth impacts every part of the local community. In hindsight, in November 2016, after the US presidential election, it should have come as no surprise that rural communities felt forgotten. For many in these areas, it seemed as if the nation's economic prosperity had made a hard stop at the border of our urban and suburban counties.

Rural counties across the country like Ferry County had helped put a populist into the White House. We had started our trip in King County, where Seattle is located and where only 22 percent voted for Donald Trump. In Ferry County, only 30 percent voted for Hillary Clinton.[11] When it comes to the nation's politics, the two counties are polar opposites. A day with time divided between the two places provides the opportunity to understand a divided nation more clearly.

This also points the way toward at least part of what it will take to create a brighter future for rural areas.

The Center for Rural Affairs understands the challenge at almost an intuitive level. Operating from its three offices in Iowa and Nebraska, it speaks with the plainspoken language of the center of the country. "We are unapologetically rural," the group says. "We stand up for the small family farmer and rancher, new business owner, and rural communities."[12]

As it turns out, the Center for Rural Affairs also has details and numbers to make the economic case for broadband adoption. Its 2018 report titled *Map to Prosperity* shows that eighty new jobs are created for every one thousand new broadband subscribers.[13] An increase of four megabits per second in residential broadband speed translates to an annual increase in household income of twenty-one hundred dollars. And people looking for work find a job 25 percent more quickly through online searches than through more traditional approaches.[14]

Today's dismal state of rural broadband in America has several causes. First and most important, installing traditional broadband and internet alternatives is expensive. Industry estimates suggest that installing fiber-optic cable—the traditional gold standard of broadband service—can cost thirty thousand dollars per mile.[15] This means that delivering sufficient broadband to remote parts of the country would cost billions of dollars, an expense the private sector has not yet been willing to pay.[16] Yet each year, the FCC's universal service mechanism and legacy programs provide eight times as much funding for landline carriers as it does for wireless carriers through its Mobility Fund and legacy programs.[17]

But this points to the second problem. Until recently, the development of alternatives to fiber-optic cable has been slow and uneven. While mobile telecommunications technologies such as 4G LTE have given customers broadband-like speed through smartphones and other mobile devices, this technology is better suited for more densely populated areas. Satellite broadband can be the right solution in very sparsely populated areas, but it often suffers from high latency, lack of significant bandwidth, and high data costs.

Third, regulatory uncertainty has contributed to challenges in bringing broadband to rural America. For example, providers seeking access to critical rights of way for network facilities often face confusing federal, state, and local permitting rules that add time and expense to projects.[18]

Finally, there is a perception that weak demand for broadband in rural areas cannot support private investment. If progress requires fiber-optic cables that generate a market return on a cost of thirty thousand dollars per mile, that's an accurate perception. But it also misses an important point. Rural demand is present and real. The market could go to work with an approach that's less costly.

This is where history and technology intersect, with an important insight for the future.

History shows that wired technologies like cable TV, electricity, and the landline telephone always take much longer to reach rural areas than wireless technologies like radio, TV, and the mobile phone. It took forty years for the landline telephone to reach 90 percent penetration, yet the cell phone reached the same threshold in just a decade. You never hear about the world needing to solve a radio or TV gap—these wireless devices were adopted quickly and were plug-and-play, latching on to the right frequency to work.[19] The lesson is obvious: If it's possible to shift from fiber-optic cables to wireless technology for broadband, we can spread broadband coverage farther and faster and at a lower cost—not just in the United States, but around the world.

Over the past decade, a new wireless technology has been emerging to fill this gap. It's called TV white spaces, and it uses the vacant channels in the TV band where signals travel a long distance. If you grew up before cable television, you either relied on a large antenna on the roof or spent time adjusting the family TV's "rabbit ears" to catch the VHF or UHF signal—the strong terrestrial signals that can travel for miles, around hills, through trees, and through the walls of our homes. Many VHF and UHF channels currently go unused and can be devoted to other purposes. And with newly developed database technology, antennae, and end-point devices, we can harness this space by connecting a TV white spaces tower to a single fiber-optic cable and rely on these wireless signals to reach towns, homes, and farms more than ten miles away.

By coincidence, I had flipped the switch to help turn on Africa's first live demonstration of TV white spaces technology. It was in

2011 at a United Nations conference in Nairobi, Kenya, and we enabled attendees to use the Xbox at broadband speeds over the Internet based on a TV white spaces signal that traveled a mile. Kenyan government officials were among the first to recognize the technology's potential, and we continued to work on it with them and several other governments. In 2015, I went back to a small rural Kenyan village on the equator, where only 12 percent of the population even had electricity. But we had partnered with a start-up to bring broadband speeds to people using TV white spaces. And I sat down and talked with teachers about rising student test scores and people who had jobs that had been unimaginable in the community just a year before.

By 2017, we concluded that TV white spaces technology was ready for broader adoption at scale, including in rural areas across the United States. After several months of planning, we launched in July what we called the Microsoft Rural Airband Initiative at the Willard InterContinental hotel in Washington, DC.

We pledged to bring broadband coverage to two million additional Americans in rural areas within five years—by July 4, 2022. We would not enter the telecommunications business, but we would partner with telecom providers and deploy a mix of wireless technologies, including new wireless devices using the TV white spaces spectrum. We pledged that for five years we would reinvest every dollar of profit from these endeavors to further expand coverage. We called for national policies to make broadband more accessible in rural areas, and we announced that we would launch twelve projects in twelve states within twelve months. And then we would grow from there.

We chose the Willard hotel for a reason—not only to attract the attention of federal lawmakers, but as a nod to a special occasion that took place at that same spot on March 7, 1916. Alexander Graham

Bell, the leaders of American Telephone &Telegraph, and luminaries from across the nation had gathered at the grand hotel for a lavish banquet hosted by the National Geographic Society to celebrate the fortieth anniversary of Bell's invention of the telephone. AT&T's leaders, however, wanted to do more than celebrate the past. They had developed a plan to use the evening to sketch a bold vision for the future.[20]

Theodore Vail, AT&T's president, wanted to inspire the nation with a vision of bringing long-distance telephones to every corner of the country, no matter how remote. It was a cause the country seized with gusto. Until that evening, people thought of commercial telephone service as something that was confined to intercity lines between the biggest cities in the country and to a few other small telephone exchanges. "Is it too much that in time it will be possible for anyone at any place to immediately communicate with anyone at any other place in the world?" Vail asked the crowd.[21]

As we know today, it was possible. And then the nation made it real. Part of our point was that the country had conquered this type of challenge before, and we were confident it could do so again.

While we committed our Airband program to bring broadband to two million people, we were clear that our real goal was much bigger. We wanted to use technology to harness the power of free enterprise and set in motion new market dynamics that would close the rural broadband gap more quickly for everyone. It meant using our funding in part to accelerate hardware innovation by chip makers and manufacturers of the end-point devices that would bring these signals into homes, offices, and farms, where they are converted to local Wi-Fi signals. It also meant bringing together small telecommunications providers into a buyer's consortium so they could purchase these devices and secure the volume discounts available only to larger purchasers.

We found we could be more targeted and move faster than any government, making progress more quickly than we anticipated. In the first seventeen months after announcing the Airband initiative, we entered into commercial partnerships in sixteen states. These partnerships will bring broadband coverage to more than one million people who lacked this access before this work began. Progress was fast enough for us to raise our ambition by the end of 2018, declaring that we would bring this coverage by 2022 not to two million people but to three million. And if additional steps are taken, it should be possible for this technology to go even faster.

Perhaps not surprisingly, Microsoft's announcements have struck a nerve. Radio talk shows and newspaper editorials in rural communities across the country have lit up with support. And we've been besieged by calls from governors and members of Congress looking for their states and districts to be added to our list.

One key to deploying this strategy is to use the right technology in the right places. We expect that TV white spaces and other fixed wireless technologies will ultimately provide the best approach to reach approximately 80 percent of the underserved rural population, particularly in areas with a population density between two and two hundred people per square mile. But other technologies, including cable-based and satellite approaches, will be needed in other areas. We believe this hybrid approach can reduce the initial capital and operating costs for the country by roughly 80 percent compared with the cost of using fiber-optic cables alone, and by approximately 50 percent compared with the cost of current LTE fixed wireless technology.

People sometimes look at us quizzically when they hear that the Airband initiative will reinvest revenue from telecommunications partnerships rather than make a profit. Why would a company spend its money this way? As we point out, the entire tech sector, including

Microsoft, will benefit when more people are connected to the cloud. In addition, we're building new applications that people can put to work in rural areas once they're connected. One of our favorites is called FarmBeats, which uses TV white spaces to connect small sensors across farmland to enable precision techniques that improve agricultural productivity and reduce environmental runoff. If we can find new ways to combine doing good with doing well, we open the door to even more investments that can reignite economic growth in rural areas.

Even with these market dynamics, however, it's important to recognize that the public sector has an important role to play in closing the broadband gap. First, we need regulatory certainty to ensure that the necessary TV white spaces spectrum remains available. While some of the TV band has been auctioned off and licensed to mobile carriers, it's important to ensure that at least two usable channels remain available to the public for TV white spaces technology in every market, with more available in rural areas. The good news is that a lot of work has already taken place and continues.

We also need public funding that is better focused in part on new technologies and not simply trying to lay expensive fiber-optic cables in the ground. Government funding can have the biggest impact at the lowest cost if it includes an opportunity for targeted funding to match money for capital investments by telecommunications companies. That's what will accelerate this work and help reach parts of the country that the private sector might be slower to reach on its own.

Ultimately, we need a national crusade to focus on and close the broadband gap. We need to recognize that, as was the case with electricity, a country separated by broadband availability will remain a nation more divided overall.

In fact, there's a lot we can learn from the steps the nation took to bring electricity beyond urban centers and to every corner of the country. Sympathetic to the rural farmer's plight, Franklin D. Roosevelt pledged in 1935 to do precisely that. He realized the country couldn't move forward into a new technological era while leaving its rural neighbors behind.

As part of his plan to lift the United States out of the economic pit of the Great Depression, FDR signed an order creating the Rural Electrification Administration as part of the nation's New Deal. The agency would help farming communities form local electrical cooperatives—a concept familiar to farmers who already bought feed and equipment through co-ops—to pay for the last miles of electrical connections. The REA's low-interest loans paid for the construction of local electrical systems that the co-op would own and oversee.

It was a program that started in Washington, DC, but its success required people who would take the promise of electricity to every corner of the country. Then as now, it required that people who wanted to change the country travel to Iowa—not to run for president, but to spread the promise of a new technology.

More than eighty years ago, the weary farmers of Jones County, Iowa, felt a similar pain to our new friends at the Knotty Pine diner. But in the summer of 1938, hope was on the horizon in the form of a glittering big top. The residents of rural Iowa had converged on the tiny town of Anamosa in eastern Iowa for the opening night of the circus—a welcome respite from a hard day and almost ten years of economic hardship.

There were no clowns, acrobats, or trained animals at this show, but the Rural Electrification Administration's traveling electric circus delighted the audience just the same. The tent featured modern-day marvels like lamps, stoves, refrigerators, poultry brooders, and milking machines, all demonstrated by the era's Vanna White: Louisan Mamer, the first lady of the REA.[22]

With a flip of a switch or a turn of a knob Louisan illuminated rooms, washed and pressed clothes, played music, swept up dust, and chilled food. At a time when cooking without electricity was a grueling affair, she made working in the kitchen look easy. She blew the crowd away as she whipped up meals of beef stew, roasted turkey, and fruit dumplings on a Westinghouse cooktop. She brought the show to a rousing close when she challenged two men from the audience to a cooking duel.[23]

When Louisan joined the REA, 90 percent of city dwellers had electricity versus 10 percent of rural Americans[24]—a gap not seen in other Western countries. At the time, electricity powered homes and barns in almost 95 percent of the French countryside.[25]

Like the large telecommunications companies today, the private electric utilities in the United States had connected the towns along the major highways but skirted the less populated areas, which were mostly farms. These companies had decided that they couldn't recoup the cost of extending their lines into far-flung swaths of rural America. And even if these rural communities were connected, the electric companies assumed American farmers, who had been particularly hard hit by the Depression, would never be able to pay for the monthly service.

This lack of electricity not only denied farmers the convenience and comfort of the modern age, it also shut them out of the nation's economic recovery. Those eager to plug into the nation's new economy had to pay private electric companies exorbitant fees to stretch lines to their land. In Pennsylvania, John Earl George was told he'd have to pay $471 to the Pennsylvania Electric Company to extend a line 1,100 feet to his home in rural Derry Township. In 1939, $471 was the average annual wage in rural Pennsylvania.[26]

In the end, REA supported 417 cooperatives across the country, serving 288,000 households,[27] and it sent Louisan and the electric circus on a four-year nationwide tour to teach farmers how to get the most from this new technology. The Maquoketa Valley Rural Electric Cooperative in Iowa was the first co-op to host the circus,[28] which by year four was drawing rural crowds of more than ten thousand people.[29]

A quarter of rural households were electrified by the end of the 1930s.[30] In Pennsylvania, John Earl George paid a five-dollar membership fee to join the Southwest Central Rural Electric Cooperative Corporation. His first bill was just $3.40.[31] When President Roosevelt died in 1945, nine out of ten farms in rural America had electricity.[32] Through public-private partnerships, persistence, and a little ingenuity, the United States had managed to shrink the rural electricity gap by 80 percent in ten years—all during an arduous economic recovery and the Second World War.

For Louisan, bringing modern technology to farmers was more than an economic imperative, it was a social cause. Raised in rural Illinois with no running water or electricity, she understood firsthand the backbreaking work of life on a farm. Lack of electricity wasn't just harming rural families' livelihoods, it was harming their lives. "I think perhaps in almost every rural home, there was a realization . . . that the drudgery of home-making in rural areas must be lightened," she said during an interview when she was in her eighties. "This heavy load of doing everything by hand the hard way, and bearing a lot of children, was killing women far earlier than they die today."[33]

As much as anything, her story is a testament to the need again to recognize that the spread of new technology is not just an economic imperative. It needs to be treated as a social cause.

As we left Ferry County, Washington, we were abuzz about everything we had seen and learned. Above all else, we talked about one question: Was there something meaningful we could do?

We didn't want to leave rural residents with a bunch of empty promises, like so many others had before. We knew our Airband initiative could help bring twenty-first-century technology to people like Elbert Koontz and his Ferry County neighbors. We asked Paul Garnett, Microsoft's Airband lead, to set out and find the right partner.

Paul and his team succeeded, and by the end of the year we announced an agreement with Declaration Networks Group to deliver broadband internet access using TV white spaces and other wireless technologies to forty-seven thousand people in eastern Ferry County and neighboring Stevens County over the next three years. It was just a start—but a real one.

In the summer of 2019—almost a year after our first visit to Ferry County—we headed back to Republic to check on progress with Declaration Networks and other new partnerships. And this time we knew the way.

As we drove out of town that evening, we made a final stop on Main Street at the Republic Brewing Company, which serves as the town hub. The front of the store has a massive garage door. When the sun's shining, the door is rolled open to let the tables spill out onto the sidewalk.

One of the owners had been tending bar when we had visited the year before. She had been surprised to hear we were from Microsoft. As we talked with her, she had laid down an opportunity and a challenge for us. "There is not a doubt in my mind that internet access around the area in the next five years will be completely different from what we see today," she mused. "There are so many bright people around here. Once they have better access to the internet, they'll realize all the different things that they can do with their lives."

It was a challenge that got us up in the morning in the months that followed. It's a challenge that needs to get the entire nation up in the morning in the years ahead.

Chapter 10 ⟫⟫ **THE TALENT GAP:**
The People Side of Technology

Most people consider technology a product business. The industry's products grab the public's attention and shape the way we work and live. But in a world where today's hits quickly become yesterday's memories, a tech company is only as good as its next product. And its next product will only be as good as the people who make it. This means, in short, that technology is fundamentally a people business.

The Fourth Industrial Revolution is defined by digital transformation. Every company in part is becoming a tech company. The same is true for governments and nonprofit groups. As a result, the people side of technology is becoming important for every slice of the economy.

The implications are multifaceted and even profound. To succeed in the digital era, companies need to recruit world-class talent, both homegrown and from elsewhere. Local communities need to ensure their citizens are equipped with new technology skills. Countries need immigration policies that give them access to the world's top talent. Employers need to develop a workforce that reflects and understands the diversity of the customers and citizens they serve. This requires not only bringing more diverse people together but also creating a culture and the processes that will enable employees constantly to learn from each other. Finally, as technology accelerates growth in key urban centers, these regions need to manage the challenges this growth is creating, not just for individual institutions, but for the entire community.

In each of these areas, tech companies are dependent on support from a community and often even a nation. And in each area, tech companies have an opportunity and a responsibility to do more themselves. It's a formidable challenge, much like a Rubik's Cube puzzle that can only be solved by moving many pieces at the same time.

How can we best advance the people side of technology?

For us, a good learning opportunity arose when we dropped by the company's annual science fair for software developers in 2018. The Microsoft Conference Center had been transformed into our annual TechFest, put on by Microsoft Research, or MSR, as everyone calls it.

MSR is one of the world's largest organizations dedicated to basic research. It's hardly typical, as it reflects an elite of the elite when it comes to people creating technology. But it provides an important initial window on the world of technology.

MSR has more than twelve hundred PhDs, eight hundred of whom have computer science degrees. To put that in perspective, the computer science departments of major universities typically employ sixty to a hundred PhDs as faculty and postdoctoral fellows. And in quality, MSR typically is considered a match for any of the top universities. Think of it as one of the world's best university computer science departments multiplied by ten in size. It's a modern-day counterpart to what AT&T assembled at Bell Labs decades ago.[1]

MSR's annual TechFest is like a trade show, but mostly open only to Microsoft employees. Teams of researchers build booths to showcase their most recent work. The goal is to enable engineers in the company's various groups to see the advances and adopt them as quickly as possible in their products.

One exhibit that was at the top of our must-see list was Private AI, a recent breakthrough that better protects people's privacy by creating the technical capability to train AI algorithms on data sets that remained encrypted. The Private AI team crowded around their exhibit and enthusiastically answered our questions. These men and women were clearly a close-knit group and knew each other well. But as the conversation wrapped up, we realized something else remarkable. This team of eight came from seven countries. There were two Americans and one person each from Finland, Israel, Armenia, India, Iran, and China. All eight now lived in the Seattle area and worked together on our Redmond campus.

This group of researchers personified something much larger than itself. Here was a team working on one of today's great technology challenges, which required a world champion lineup—one that America's immigration system had enabled us to bring together.

The immigration issue has long been a tough one for the tech sector in the United States. At one level, it has been indispensable to the country's technology leadership in the world. There is simply no way the United States would be the global leader in information technology if it had not attracted many of the best and brightest people in the world to come work at leading universities or live in technology centers around the country.

Immigration's role in innovation was important to the United States when the country's West Coast economy was still dominated by agriculture, and silicon was associated only with sand. The country's ability to attract Albert Einstein from Germany at the height of the Great Depression played a vital role in awakening President Franklin Roosevelt to the need to create the Manhattan Project.[2] Its open door to German rocket scientists after World War II was critical to sending the first man to the moon. With the help of federal investments in basic research at the country's great universities and President Eisenhower's support of math and sciences in the nation's public schools,[3] the United States developed an approach to research, education, and immigration that led to decades of global economic and intellectual leadership.

The rest of the world studied this model and increasingly emulated it. But Americans increasingly forgot what made it work. And the political support for its various pieces began to fall apart.

The tech sector confronted this growing discord as it dealt with immigration challenges soon after the twenty-first century began. Year after year, the Republicans would support highly skilled immigrants but not broader immigration reform. The Democrats would support highly skilled immigrants but only as part of broader immigration reform. Years of talking with leaders of both parties almost always ended with the frustrating conclusion that nothing would get done. After the 2016 presidential race, it only became worse.

As Satya and I flew to New York in December 2016 for President-Elect Trump's meeting with tech leaders at Trump Tower, we decided that we would find a way to raise the issue of immigration at some point in the conversation. Satya remarked early in the discussion on its importance in his personal life and its continuing importance today. No one else weighed in until Trump asked the group in a gracious way whether we wanted to talk about our views. We dived into important details. As we did so, he said we had nothing to worry about. "Only the bad people will have to leave," he said. "All the good people can stay and continue to come." Who could disagree with that? But who knew what it really meant?

We talked on the sidelines of the meeting with the incoming White House staff about immigration and education issues. It pro-

vided some hope. But in February 2017, a month after the inaugura-
tion, this hope evaporated. The new president issued a complete
travel ban on individuals from seven Muslim nations. Across the nation, people gathered at airports to protest the targeting of countries
based on religion. At Microsoft, the ban impacted 140 Microsoft employees and family members, including a dozen who happened to be
outside the country and could not return.

Across the tech sector, there was no doubt in our minds on which
side we stood. We stood with our people. We had employees and
families at risk, and we would see them through this crisis.

Within hours, Washington state attorney general Bob Ferguson
decided to file suit. Amazon's thoughtful general counsel, David
Zapolsky, was especially instrumental in the first days, as we sought to
define a strategic path.[4] I organized a call the next Sunday afternoon,
and with Apple, Amazon, Facebook, and Google, we decided we'd
work together to organize broad tech sector support in a legal brief.

Despite the drama unleashed by the travel ban, we hoped the
issue might settle down and that room for compromise might emerge.
Satya and I traveled to the White House in June 2017 for another
meeting involving tech leaders. The comprehensive set of meetings
was planned by Chris Liddell, the former Microsoft CFO who was
spearheading under Jared Kushner a series of initiatives to modernize the federal government. I participated in a frank and wide-ranging
breakout discussion that explored whether there could be a broader
immigration package. While there was a clear schism among different parts of the White House staff, we had some renewed hope.

But by the beginning of September it was clear that the White
House—and the nation—was approaching the next immigration Rubicon. The question was whether the president would pull the plug
on the Deferred Action for Childhood Arrivals, or DACA program,
throwing the lives of more than eight hundred thousand young

DREAMers, including some of our employees, into complete limbo. We urged a compromise that would address border security while keeping DACA and other key immigration measures intact.

It all came to naught. As I talked with people at the White House in the final hours before a decision was to be announced, the situation looked increasingly bleak. I brainstormed with Amy Hood, our CFO, about what we could do to protect our DACA employees. We developed a plan and Satya signed off on it. When the president announced the decision to rescind DACA, we were ready. Microsoft became the first company to commit to providing legal defense for affected employees. As I said to an NPR reporter, if the federal government wanted to deport any of our DACA employees, "it would have to go through us."[5] We then joined with Princeton University and one of its students to file a lawsuit to challenge the DACA rescission.[6]

In many respects, the DACA decision set the terms for each subsequent immigration discussion. Talk of compromise would emerge and then fall apart. It was also, however, part of a pattern that had persisted for a decade. President George W. Bush tried to break the immigration deadlock with comprehensive legislation in his second term. President Obama tried in his second term as well, with the Senate passing a comprehensive bill in 2013. But ultimately gridlock was the only winner.

Now, however, the debate was even more shrill. Each party easily found its way back to its political corner. Each would double down on appeals to its base, which were not difficult to fashion. The only thing that suffered was the opportunity to get something—anything—done.

Away from the public glare of politics, we sometimes encounter a similar situation. A tug-of-war emerges over a single issue in the world of business or regulation. It becomes a contest that inevitably will produce one winner and one loser. It's a recipe for a sustained impasse, for getting nothing done.

Ironically, the answer to such problems is sometimes to broaden the challenge. One tenet I always employed in negotiations was a simple one: Never let a negotiation narrow to a single issue that can produce only one winner, even if it means holding open some other topics on which agreement might seem in reach. Instead, broaden the discussion so that more issues are on the table. Create the opportunity for more give-and-take and a round of compromises that can produce a scenario that enables everyone to claim victory in the final stage. It was an approach that had proven indispensable to our ability to work through difficult antitrust and intellectual property disputes with governments and companies around the world.

It led us to believe that it could play a role in addressing immigration issues as well. After all, there needed to be a real and fair balance between filling new tech jobs in the United States with immigrants and creating even more jobs that would be filled by American citizens.

This was an issue of both principle and pragmatic politics. We had spent enough time working on immigration to appreciate that the biggest political challenge was the threat people perceived it posed to opportunities for those born in the United States. It was something we also saw in many other countries where we have employees. As with trade, immigration advances can be perceived as threats to jobs for the domestic population. But immigration often is even more politically controversial because it can be perceived as upsetting a domestic culture as well, given the influx of people and customs from other countries.

We had proposed what felt like our best idea in 2010, when we advocated for a "national talent strategy" for the United States.[7] We sought to broaden the issue and advance immigration in a way that would also create more opportunities for Americans. The notion was to couple a limited increase in the supply of visas and green cards with a larger hike in immigration fees, using the added revenue to fund broader education and training opportunities for the skills most in demand for new jobs.

There naturally was a slew of details to work through. A group of senators took on this challenge in 2013, when Orrin Hatch and Amy Klobuchar led a bipartisan effort that introduced what they called the Immigration Innovation Act.[8] Known as I-Squared, the bill adopted the basic formula we had proposed while also addressing the critical shortage of green cards for certain key countries and making other overdue reforms. Most of its provisions made their way into the comprehensive immigration bill passed by the Senate in 2013, only to languish in the House of Representatives. When we discussed immigration at Trump Tower in December 2016, I raised the approach again. Most of the tech leaders were supportive, but the president-elect's staff was clearly divided.

Part of the attraction of I-Squared was that it would raise money for a cause that clearly has grown in importance. Every nation confronts a new imperative to make it easier for people to develop the skills they need for better job opportunities in an AI- and technology-based economy. It's an issue important for us to address head-on as a technology company as part of our own hiring. And the antitrust legal difficulties we faced in the 1990s had led us to become more deeply involved in a way that provided added insight into the issues.

One of our own turning points came in early January 2003, just as everyone returned from the holidays. Our litigation team had hammered out an agreement in principle to settle what we knew would be the single biggest class-action lawsuit to result from our antitrust loss before the federal court of appeals in Washington, DC. It covered all the consumers in California. The price tag was a whopping $1.1 billion. It would be the largest litigation settlement in Microsoft's history. I sent Steve Ballmer, then Microsoft's CEO, an email to let him know that I wanted to move forward, and I held my breath waiting for his reaction.

Steve walked down the hall and into my office on that same morning to talk about the proposed settlement. He understood, as almost all business executives do, that the lawyers who bring class action lawsuits always ensure that they do well for themselves as part of these settlements. But Steve wondered what else this one involved. He paced around my office, as he often did. He then sat down, but not in a chair. Instead, he sat on top of my desk cross-legged— something I hadn't seen before. He looked me straight in the eye and said, "If we're going to spend all of this money, I want you to ensure that some real people get some real benefit from this." I promised that I would.

The final settlement fulfilled Steve's request. Microsoft agreed to provide vouchers to schools so they could buy new computer technology. Not just *our* technology, but software, hardware, and services from any company, including our competitors. It provided a model that we would follow across the country, ultimately providing more than $3 billion in vouchers for schools nationwide.

Over time, however, this settlement gave us a firsthand insight into what the rest of the country also learned: Despite spending billions of dollars, the biggest technology challenge for schools was not getting more computers into classrooms. It was equipping teachers with the skills needed to put that technology to use. And little did we know that the biggest skill challenge for teachers was yet to come: providing the opportunity for them to learn computer science, a field that was only in its infancy when many of them went to high school or college. Only then could they teach the coding and computer science courses that would define the future for a new generation of students.

Computer science has rapidly become one of the defining fields of the twenty-first century. Jobs have progressively become more digital in their content, and as a Brookings Institution study concluded in 2017,[9] jobs with more digital content pay better than jobs with less.[10] As leading professor Ed Lazowska has noted at the University of Washington, computer science has become "central to everything." As he explains, "It's not just software, it's biology, you name it, computer science is at the center of it."[11]

But there's a huge shortage of teachers who can teach computer science. Less than 20 percent of the nation's high schools offer the Advanced Placement course in the field.[12] The number of young people taking the AP course in 2017 was lower than for fifteen other subjects, including European history. One challenge is the high cost of training teachers so they can teach computer science.[13]

While governments have been slow to address the problem, philanthropy has moved faster. One individual who has made a difference is Kevin Wang. He had graduated with both computer science and education degrees and had taught high school before becoming a software engineer. Three years into his career at Microsoft, a local Seattle high school learned about his background and asked if he would volunteer to teach a computer science class. He did, and soon other local schools were asking if he could teach there too.

Kevin explained that he had a day job and obviously couldn't teach in five places at once anyway. But if they were interested, he knew other Microsoft developers who could do a good job team-teaching with someone like the school's math teacher. The volunteer would provide expertise in computer science, while the math teacher knew how to teach, manage a classroom, and work successfully with students. Working with the volunteer, the math teacher over time would also learn computer science and become a computer science teacher as well. A new approach to teacher training was born.

Microsoft Philanthropies made the new program—Technology Education and Literacy in Schools, or TEALS—a cornerstone of its educational mission. It annually enlists 1,450 volunteers from Microsoft and 500 other companies and organizations to teach computer science in almost 500 high schools in 27 US states, plus the District of Columbia and British Columbia, Canada.

A second individual then stepped forward, and he has had an even bigger impact. Hadi Partovi had successfully founded and funded a variety of new tech companies up and down the West Coast. He is the son of Iranian parents who fled their country and emigrated to the United States in the wake of the Iranian revolution. Despite Hadi's success, his father wondered when he would accomplish something even more important. Hadi's response was to take some of his own money to found a new group, Code.org, that would change the face of computer science education.[14]

From the perspective of a traditional nonprofit, the reach of Code .org is awe-inspiring. To introduce a new generation to coding, Hadi created an annual program called Hour of Code, inspiring students to try coding through one-hour online tutorials. He put his viral marketing skills to work, and to date hundreds of millions of students have participated globally.[15] Microsoft became Code.org's largest funder, and we've cheered the organization as it has expanded teacher training and support across the country.

The problem is that a lot more support is still needed to better reach students from all backgrounds. While you would be hard-pressed to say that every student must take computer science, you could say that every student deserves the opportunity. That means getting computer science into every high school, and into earlier grades as well. The only way to train teachers at this scale is for federal funding to help fill the gap.

After years of lobbying, there was a breakthrough in federal interest in 2016. In January President Obama announced a bold proposal to invest $4 billion of federal money to bring computer science to the nation's schools. While the proposal produced enthusiasm, it didn't spur Congress to appropriate any new money.[16]

Ivanka Trump had more success the following year. Even before her father had moved into the White House, she was interested in federal investments in computer science in schools. She was confident she could persuade the president to support the idea, but she also believed that the key to public money was to secure substantial private funding from major technology companies. She said she would work to secure $1 billion of federal support over five years if the tech sector would pledge $300 million during the same time.

As always, there was the question of whether someone would go first. The White House was looking for a company to get things rolling by pledging $50 million over five years. Given Microsoft's long-standing involvement, financial support, and prior advocacy with the Obama White House, we were a natural choice. We agreed to make the commitment, other companies followed, and in September 2017 Mary Snapp, the head of Microsoft Philanthropies, joined Ivanka in Detroit to make the announcement.

The need for computer science in the nation's schools is a gating factor for opportunities for a new generation in an evolving economy. But it's just part of the equation. Increasingly, nonprofit groups and state governments are developing innovative programs to strengthen local schools, invest in community colleges, improve lifelong learning, and explore new career pathways for individuals who will need to change jobs or careers as their lives progress. Groups from across the country are traveling abroad, asking whether the Swiss apprenticeship model or Singapore's lifelong learning financial accounts can be adopted successfully in the United States. It's a national challenge, and away from the gridlock in Washington, DC, progress is sprouting across the country.

The tech sector is also investing in learning and job-seeking tools, including Microsoft's own work at LinkedIn. LinkedIn created the Economic Graph,[17] which identifies what types of jobs companies are creating by region and country and what types of skills are needed to fill those jobs. Informed by data from more than six hundred million members around the world, the Economic Graph provides a tool to help public planners focus their education and skilling programs. From Colorado to Australia to the World Bank, it is being used by governmental and nonprofit organizations alike.[18]

As LinkedIn's data makes clear, skills that build on computer and data science have become increasingly important to new job seekers. In May 2019, the top four skills pursued by recent college graduates on LinkedIn Learning—data visualization, data modeling, programming languages, and web analytics—all reflected this emphasis.[19] Microsoft's top two sales leaders, Jean-Philippe Courtois and Judson Althoff, recognized that the adoption of new technology increasingly requires that the company invest in skills-development programs not only for our own employees but for the employees of our customers. This has led to the creation of a program that will bring AI and other technology skills to customers around the world.

As we make progress, new lessons and challenges continue to emerge. One ongoing challenge is making sure that people can acquire new skills in an affordable way, including by moving from computer science classes in high school to a college degree or other postsecondary credential. It's a challenge that the tech sector and governments can address together by partnering in new ways.

We've experienced this firsthand in Washington, where the Washington State Opportunity Scholarship program, created by the state legislature, matches public funds with private dollars to help local students pursue a college degree in health care or in science, technology, engineering, and math.[20] Since 2011 this combination has raised almost $200 million, supporting roughly 5,000 college students a year, each of whom can receive up to $22,500 in scholarship money. The program has broadened access to college, with nearly two-thirds of the recipients being the first in their family to go to college and a majority being women and students of color.[21]

While all this has been good news for major private-sector funders and employers such as Microsoft and Boeing, what's even more encouraging is the program's broader approach and the outcomes it has produced. When executive director Naria Santa Lucia joined the effort five years ago, she focused on providing students with mentors, internships, and connections to potential employers. This has created roles for businesses and individuals across the community. The combination has not only led to strong graduation rates for the students, it has provided a clear path into well-paying jobs. A recent review found that just five years after graduation, the median income of participants was almost 50 percent higher than that of their entire family at the time their college studies began. This is at a time when, more broadly across the country, the odds of thirty-year-old Americans earning more than their parents at the same age "has fallen from 86 percent forty years ago to 51 percent today."[22]

This success inspired us to pursue an even bigger local effort to broaden access to new skills and higher education. In early 2019, local leaders asked whether I would join University of Washington President Ana Mari Cauce in encouraging the creation of a new education fund that would be paid through an increased tax on an array of businesses that rely on the higher education system.

While the idea was appealing, it also raised its share of challenges. I wanted to ensure that the funds would reach not only those pursuing degrees at four-year institutions but also help students attending technical schools and community colleges. I also wanted to make sure that there would be an independent board to evaluate whether the money was being spent responsibly, plus provisions that would safeguard against the funds being diverted during a recession.

While these issues proved relatively easy to address, it still felt uncomfortable to advocate that other businesses pay more in taxes. Amy Hood and I sat down to think through what Microsoft's advocacy would mean for the company both financially and politically. We concluded that if we were going to put the company's reputation on the line advocating for an additional tax on local businesses, we should propose a structure that would require Microsoft and Amazon, as the two largest tech companies in the state, to pay a higher tax rate than everyone else.

That's what we proposed, and as it turned out, that's what the legislature enacted. We kicked off our public advocacy in an op-ed in the *Seattle Times*,[23] proposing a surcharge on the state's business and occupation tax on services. As we wrote in the op-ed, "Let's ask the largest companies in the tech sector, which are the largest employers of high-skilled talent, to do a bit more."[24] While the proposal initially raised some friction with other companies,[25] legislators found their way to a compromise that capped the new tax surcharge at $7 million per company per year. Just six weeks later, the legislature approved a new budget with a dedicated fund that will raise roughly $250 million per year for higher education.

Washington's new Workforce Education Investment Act was hailed both locally and nationally for its commitment to "provide free or reduced tuition for lower- and middle-income students attending community colleges and public institutions, provide new funding for strapped community colleges and eliminate wait lists for financial aid beginning in 2020."[26] As described by a professor at Temple University, it was "pretty much the most progressive state higher ed funding bill" enacted in several years.[27] For me, it was proof that if the tech sector could get comfortable with a more community-focused approach and pay perhaps even a bit more than its share toward it, we could have a real and positive impact.

Unfortunately, this type of progress remains a bit of an oasis in the desert. In the United States, access to technology skills is far from distributed evenly. Like the broadband gap, the skills gap hits some groups far harder than others. It's exacerbating almost every other divide that afflicts America.

You see the disparate impact clearly when you look at the students learning computer science. At a time when technology suffers from a shortage of women, only 28 percent of the students taking an AP computer science exam in 2018 were girls.[28] You see the same trend play out with underrepresented racial minorities. These groups comprised just 21 percent of the students taking these exams, compared to their 43 percent representation in the country.[29] And at a time when the country worries about economic opportunities in rural communities, only 10 percent of the students who took AP computer science exams in 2018 were from these communities.[30]

In short, the students taking the AP computer science class are more male, more white, more affluent, and more urban than the country as a whole. This part of the problem has multiple causes. But the tech sector needs to accept its share of responsibility. It has not always been an easy place for women or minorities to build a career.

Science and technology have long had prominent female pioneers, including Marie Curie, who remains the only person to twice win the same Nobel Prize, and Bertha Benz, the first person to show the world the automobile's potential.[31] But while men were prepared to acknowledge the contributions of these women as individuals, the world of technology remained stubbornly slow in recognizing and creating opportunities for women more broadly. At most tech companies, women still represent less than 30 percent of the workforce, and an even lower percentage of technical roles. Similarly, African

Americans, Hispanics, and Latinos typically account for less than half of what one would expect based on their representation in the American population.

Thankfully, in the past few years this view has finally started to change. Across the sector, tech companies have launched new programs to pursue more diverse hiring and foster a more inclusive culture in the workplace. Some of the new advances have come from applying basic business practices that many other economic sectors have long embraced. Like basing some of the pay of senior executives on whether there has been real progress in advancing diversity rather than simply talking about the problem. Or deploying a group of recruiters to help identify strong diverse candidates and redoubling efforts to visit historically black colleges and the universities that have large and successful Hispanic populations.

It's not rocket science. It's not even computer science. It's mostly common sense. The good news is that the wheels have finally started to turn. But without question, when it comes to inclusion the tech sector has much more progress ahead of it than behind it.

Perhaps more of an outward focus can help tech companies think more broadly about a final dimension of the people equation as well. This is the impact they are having on the communities where they have been growing so quickly.

Fast-growing tech companies bring high-paying jobs to a community. What region wouldn't want that? The competition to attract Amazon's HQ2 revealed this dimension more than ever. City after city fell over itself to court the company and its request for tax breaks and other incentives.

But growth brings its own challenges. While it's a high-class problem to have, it nonetheless is a problem that needs to be addressed. And in many regions, things have been getting worse.

The first place you see the problem is on the highway. Traffic slows, commutes lengthen, and tech companies start providing bus services for their employees. Most weekday afternoons in Silicon Valley, the highway feels like a parking lot—except that you can drive a bit more quickly through a parking lot. The strains on highways are like the visible part of the iceberg. It's the easiest to see, but growth places the same demands on every part of a region's infrastructure, from transit systems to schools.

Over the past few years, the problem has reached a much deeper level. While jobs have grown, the housing supply has not kept pace. Basic economics goes to work. When people move in to take higher-paying jobs and housing construction fails to keep pace, housing prices rise and many low- and middle-income people are forced to move out. The community's teachers, nurses, and first responders—as well as the support staff at tech companies themselves—are often pushed into more remote areas, where they endure ever longer commutes.

Satya and I found ourselves talking about this issue at a small meeting in Seattle in June 2018. For years, we had encouraged local business leaders to focus on education and transportation as foundational issues for the entire region. Now we were having breakfast with ten other local leaders as part of Challenge Seattle, a local civic and business group we had helped to found that is led by former Washington governor Christine Gregoire, its CEO. The question that morning was about the group's priorities for the future.

The breakfast brought an epiphany. As the discussion circled the table, every attendee talked about how the region was changing in ways that were not altogether positive. In Seattle, we had long prided ourselves on avoiding the worst of the housing challenges that impacted San Francisco and Northern California. Until we realized this was no longer the case. As companies like Amazon and Microsoft continued to grow, we were joined by burgeoning engineering outposts from more than eighty companies based in the Silicon Valley.

Suddenly the Seattle area had evolved from the Emerald City to Cloud City. Between 2011 and 2018, median home prices increased by 96 percent, while median household income rose by only 34 percent.[32]

The issues hit home earlier that year in downtown Seattle. Plagued by persistent and increasing homelessness, Seattle's city council had responded by proposing to raise $75 million annually to address the problem through a head tax on jobs.[33] The business community reacted loudly with frustration, and Amazon halted construction planning for a new Seattle tower, threatening to slow job growth if the decision was not reversed.[34] We saw the issue unfold from the other side of Lake Washington, which separates Seattle from many of the other cities in the region, including Redmond, where Microsoft is based. While not involved in the Seattle debate, we watched with mixed feelings. We shared the skepticism about taxing jobs but felt that the business community needed to do more than criticize the measure. It needed to step up in a bigger way. Seattle's mayor and city council canceled the head tax, but little emerged in terms of an effective alternative.[35]

At the breakfast, Satya raised housing concerns and was quickly joined by others. I mentioned that I had sat down for a cup of coffee on a recent Saturday morning with Steve Mylett, the police chief of Bellevue, the largest city outside Seattle. I had asked to meet to share concerns raised by some of our employees about racial challenges they sometimes face in the community, including their perceptions of local police. He was open and receptive to hearing my views, and he shared a fact that was new to me: The increase in housing prices meant that new Bellevue police officers could no longer afford to buy a home in the city they patrolled. Even the chief of police endured a commute of an hour each way to work. There was an important connection between our two points: It's difficult to build a stronger connection between a community and its police force when local officers can't afford to live close by.

I shared the story with the group from Challenge Seattle and mentioned that I had asked our team at Microsoft to develop ideas for a new initiative. As Satya and I walked out of the breakfast together, I described them in more detail. By the time we reached the elevator, we had decided to make the initiative a priority.

Back in Redmond, we put a data science team to work so we could better understand the problem. The team collaborated with Zillow to include real estate data, creating a bigger data set than had been available before. The insights were eye-opening, not only for us but ultimately for the entire region. The data showed that we had not just a homelessness problem, but a rapidly expanding crisis for affordable housing more broadly. Jobs in the region had grown 21 percent, while growth in housing construction had lagged at 13 percent.[36] The gap was even bigger in the smaller cities outside Seattle, where construction of both low- and middle-income housing had stagnated. People in low- and middle-income families increasingly were being forced into towns and suburbs much farther from their jobs. The region now ranked among the worst in the country for the percentage of people enduring daily commutes in excess of ninety minutes.[37]

We decided that something needed to be done to increase the supply of low- and middle-income housing. We spent months consulting with people and groups across the region and learning everything we could across the country and around the world. With Satya's support, Amy Hood and I decided to sponsor a bigger project internally, and she put her finance team to work on developing alternatives. We quickly recognized that Microsoft, like some other large tech companies, was in the fortunate position of having a powerful balance sheet with liquid assets we could put to work. In January 2019,

Amy and I announced that Microsoft would commit $500 million in a combination of loans, investments, and philanthropic donations to address the issue.[38]

Two insights emerged from our work that seemed especially important. First, it was readily apparent that money alone could never solve the problem. As we studied the issue around the world, it became clear that the only effective path to progress was to combine more capital with public policy initiatives. As important as our funding was the simultaneous announcement made by the mayors of nine local cities to consider reforms to grow the low- and middle-income housing supply. In the run-up to our announcement, Christine Gregoire had hammered out with the mayors a set of specific recommendations to donate public lands, adjust zoning requirements, and make other changes to accelerate new construction. The issues were thorny, and these steps required more than a little political courage.[39] We hoped that as much as anything else, our funding could serve as a catalyst for the broader effort needed to bring the community together.[40]

The second insight emerged from the reaction to our commitment. It was quickly apparent that the issue struck a nerve, not just locally but nationally and even internationally. The 2016 presidential election results had reflected a broadening concern in many rural communities that in an era defined by tech-fueled prosperity, people in these areas were being left behind. But now, this concern was spreading in new ways into urban areas as well. People might walk down a city street in the shadow of new and glistening towers built to house tech sector employees, but they could no longer afford to live nearby themselves.

This was creating understandable frustration that may well be adding a new dimension to American politics. We quickly saw it spill over in New York City, where some local politicians had more than the ordinary buyer's remorse over their success in attracting Amazon's jobs through subsidies and tax breaks. We couldn't help but appreciate the issue, given the impact of the company's growth on our own region's housing needs.

In some respects, the affordable housing issue underscores the interconnected nature of all the people issues relevant to the tech sector. To build a healthy business, it's critical to have a diverse employee population and be part of a thriving community. While it's reasonable for tech companies to ask what their communities will do for them, the industry has reached the point where it needs to ask itself a larger question. Success brings with it not just size but responsibility. And the tech sector increasingly needs to ask itself what more it will do to support the communities in which it lives. We need access to great talent, not just from across the street but from around the world. But we need to do more to foster opportunity for the people all around us.

All these challenges require action. As I sometimes say at Microsoft when we start a new project, first prize is to do something big. Second prize is to do something.

Success rarely comes to people who do nothing.

Chapter 11

AI AND ETHICS: Don't Ask What Computers Can Do, Ask What They Should Do

When I arrived in Davos, Switzerland, at the World Economic Forum in January 2017 for the annual bellwether event discussing global trends, AI was the talk of the town. Every tech company was touting itself as an AI company. One night after dinner, I called on my northeastern Wisconsin roots and braved the snow and ice to walk the full two-mile length of Davos's main boulevard. It looked more like the Las Vegas strip than an Alpine village. Other than a handful of banks, the ski town was dominated by the lit-up logos and slick signage of tech companies, each (Microsoft included) promoting their AI strategy to the business, government, and thought leaders spending the week in the Swiss Alps. Two things were abundantly clear: AI was the new thing, and tech companies have big marketing budgets.

After sitting through numerous discussions about the benefits of AI, I realized that no one was taking time to explain what AI is or how it works. It was assumed that everyone in the room already knew. From my own conversations at Davos, I knew this wasn't the case, but people were understandably reluctant to raise their hands and ask about the basics. No one wanted to be the first to admit that they (and most likely half the audience) didn't fully understand what the other half of the audience was talking about.

In addition to the universal vagueness swirling around AI, I noticed something else. No one wanted to talk about whether this new technology would need to be regulated.

During a webcast on AI hosted by David Kirkpatrick of Techonomy, I was asked whether Microsoft thought there would be government regulation of AI. I said that in five years we'd probably find ourselves debating government proposals focused on new AI regulations. An executive from IBM disagreed, saying, "You can't predict the future. I don't know if we can have precise policy. I'd be worried about that having adverse effects."[1]

The week in Davos captured the themes that were pervasive in the tech sector, not necessarily all positive. The industry, like most, often rushes forward with an innovation without helping people understand fully what it is or how it works. This is coupled with what for too long was an almost theological belief that new technology will be entirely beneficial. Many in Silicon Valley long believed that government regulators could not keep up with technology.

While this idealistic view of technology was often rooted in good intentions, it's just not realistic. Even the best technologies have unintended consequences, and the benefits are seldom spread uniformly. And this is before the new technology is misused for harmful ends, as it inevitably will be.

In the 1700s, soon after Ben Franklin created the postal service in the United States, criminals invented mail fraud. In the 1800s, with the telegraph and the telephone, criminals invented wire fraud. In the twentieth century, when technologists invented the internet, it was apparent to anyone who knew history that the invention of new forms of fraud was unavoidable.

The challenge was that the tech sector, to its credit, always looked forward. The problem was that, to its detriment, too few people spent time or even accepted the virtue of looking in the rearview mirror long enough to use a knowledge of the past to anticipate the problems around the corner.

Within a year of the AI party in Davos, artificial intelligence started to create a broader set of questions for society. The public's trust in technology previously had centered around privacy and security, but artificial intelligence was also now making people feel uneasy and was quickly becoming a central topic of public discussion.

Computers were becoming endowed with the ability to learn and make decisions, increasingly free from human intervention. But how would they make these decisions? Would they reflect the best of humanity? Or something much less inspiring? It had become increasingly apparent that AI technologies desperately needed to be guided by strong ethical principles if they were to serve society well.

This day had long been in the making. Several years before researchers at Dartmouth College held a summer study in 1956 to explore the development of computers that could learn—marked by some as the birth of academic discussion about AI—Isaac Asimov had written his famous "three laws of robotics" in the short story "Runaround."[2] It was a science fiction account about humanity's attempt to create ethical rules that would guide the autonomous AI-based decision making of robots. As dramatically illustrated in the 2004 film *I, Robot* starring Will Smith, it did not go well.

AI has developed in fits and starts since the late 1950s, most notably for a short time in the mid-1980s with a flurry of hype, investment, start-ups, and media interest in "expert systems."[3] But why did it burst onto the scene in such a big way in 2017, sixty years later? It's not because it was a fad. To the contrary, it reflected trends and issues that were far broader and had long been converging.

There is no universally agreed-upon definition of AI across the tech sector, and technologists understandably advance their own perspective with vigor. In 2016, I spent some time on the emergence of new AI issues with Microsoft's Dave Heiner, who at the time was working with Eric Horvitz, who had long led much of our basic research in the field. When I pressed Dave, he provided me with what I still regard as one helpful way to think about AI: "AI is a computer system that can learn from experience by discerning patterns in data fed to it and thereby make decisions." Eric uses a somewhat broader definition, suggesting that "AI is the study of computational mechanisms underlying thought and intelligent behavior." While this often involves data, it can also be based on experiences such as playing games, understanding natural languages, and the like. The ability of a computer to learn from data and experience and make decisions—the essence of these definitions of AI—is based on two fundamental technological capabilities: human perception and human cognition.

Human perception is the ability of computers to *perceive* what is happening in the world the way humans do through sight and sound. At one level, machines have been able to "see" the world since the camera was invented in the 1830s. But it always took a human to understand what was depicted in a photograph. Similarly, machines have been able to hear since Thomas Edison invented the phonograph in 1877. But no machine could understand and transcribe as accurately as a human being.

Vision and speech recognition have long been among the holy grails for researchers in computer science. In 1995, when Bill Gates founded Microsoft Research, one of the first goals of Nathan Myhrvold, who headed the effort, was to recruit the top academics in vision and speech recognition. I still recall when Microsoft's basic research team optimistically predicted in the 1990s that a computer would soon be able to understand speech as well as a human being.

The optimism of Microsoft researchers was shared by experts across academia and the tech sector. The reality was that speech recognition took longer to improve than experts had predicted. The goal for both vision and speech was to enable computers to perceive the world with an accuracy rate that would match that of human beings, which is less than 100 percent. We all make mistakes, including in our ability to discern what others are saying to us. Experts estimate that our accuracy rate for understanding speech is about 96 percent, with our brain filling in the gaps so quickly that we don't think about it.[4] But until AI systems match this level, we're more likely to be bothered by the mistakes computers make than to be impressed by something like a 90 percent success rate.

By the year 2000, computers had reached the 90 percent threshold in vision and speech, but progress stalled for a decade. After 2010, progress picked up again. When people a hundred years from now look back at the history of the twenty-first century, they'll likely conclude that the decade from 2010 to 2020 was when AI came together.

Three recent technological advances have provided the launchpad that has let AI take flight. First, computing power finally advanced to the level needed to perform the massive number of calculations needed. Second, cloud computing made large amounts of this power and storage capacity available to people and organizations without the need to make large capital investments in massive amounts of hardware. And finally, the explosion of digital data made it possible to build massively larger data sets to train AI-based systems. Without these building blocks, it's doubtful that AI would have accelerated so quickly.

But it took a fourth foundational element, which has been critical in helping computer and data scientists to make artificial intelligence effective. This goes to the second and even more fundamental technological capability needed for AI, cognition—in other words, the ability of a computer to reason and learn.

For decades, there was a lively debate about the best technological approach to enable computers to think. One approach was based on what's called "expert systems." Especially popular in the late 1970s and 1980s, this involved the collection of large amounts of facts and the creation of rules that computers could apply in chains of logical reasoning to make decisions. As one technologist has noted, this rule-based approach couldn't scale to match the complexity of real-world problems. "In complex domains, the number of rules can be enormous, and as new facts are added by hand, keeping track of exceptions to and interactions with other rules becomes impractical."[5] In many ways, we go about our human lives not by reasoning through rules, but by recognizing patterns based on experience.[6] In hindsight, a system based on such detailed rules was perhaps an approach that only lawyers could love.

Since the 1980s, an alternative approach to AI has proved superior. This approach uses statistical methods for pattern recognition, prediction, and reasoning, in effect building systems through algorithms that learn from data. During the last decade, leaps in computer and data science have led to the expanded use of so-called deep learning, or with neural networks. Our human brains contain neurons with synaptic connections that make possible our ability to discern patterns in the world around us.[7] Computer-based neural networks contain computational units referred to as neurons, and they're connected artificially so that AI systems can reason.[8] In essence, the deep learning approach feeds huge amounts of relevant data to train a computer to recognize a pattern, using many layers of these artificial neurons. It's a process that is both computationally and data intensive, which is why progress required the other advances previously noted. It also required new breakthroughs in techniques needed to train multilayer neural networks,[9] which started to come to fruition about a decade ago.[10]

The collective impact of these changes led to rapid and impressive advances in AI-based systems. In 2016, the team at Microsoft Research's vision recognition system matched human performance in a specific challenge to identify a large number of objects in a library called ImageNet. They then did the same thing with speech recognition in a specific challenge called the Switchboard data set, achieving a 94.1 percent accuracy rate.[11] In other words, computers were starting to perceive the world as well as human beings. The same phenomenon happened with translation of languages, which requires in part that computers understand the meaning of different words, including nuance and slang.

Quickly, the public began to worry when articles appeared asking if an AI-based computer could think fully by itself and reason at superhuman speeds, leading to machines that could take over the world. It's what technologists call superintelligence or, as some put it, a "singularity."[12] As Dave Heiner argued when we focused on the question in 2016, the issue was taking up too much time and attention—arguably distracting from more important, immediate issues. "This of course is very sci-fi, and it's obscuring the more immediate questions that AI is starting to create," he said.

Those more immediate issues had come up that same year at a conference sponsored by the White House. The talk of the conference had been an article published in *ProPublica* titled "Machine Bias."[13] The subtitle for the piece said it all: "There's software used across the country to predict future criminals. And it's biased against blacks." AI was increasingly being used to make predictions in a wide variety of settings, and concern was mounting over whether the systems were biased against certain groups in various scenarios, including people of color.[14]

The bias problem described by *ProPublica* in 2016 was a real one. It reflected two real-world causes that each need to be addressed if AI is to perform the way the public rightfully expects. The first involves work that includes biased data sets. For example, a facial recognition data set with photographs of people's faces might have enough photos of white males to predict at a high accuracy rate the faces of white men. But if there are smaller data sets of photos of women or people of color, then a higher error rate for these groups is likely.

That in fact is precisely what two PhD students discovered through their research on a project called "Gender Shades."[15] MIT researcher Joy Buolamwini, a Rhodes scholar and poet, and Stanford University researcher Timnet Gebru undertook work to advance public understanding of AI bias by comparing the facial-recognition accuracy rates for people of different genders and races. The two women documented higher error rates, for example, in identifying gender for the faces of black politicians in Africa compared to white politicians in northern Europe. As an African American woman, Buolamwini even found that some systems identified her as a man.

Buolamwini and Gebru's work helped reveal a second dimension of bias that we also need to consider. It's difficult to build technology that serves the world without first building a team that reflects the diversity of the world. As they found, a more diverse group of researchers and engineers is more likely to recognize and think harder about bias problems that may even impact them personally.

If AI imbues computers with the ability to learn from experience and make decisions, what type of experiences do we want them to have and what decisions are we comfortable having them make?

At Microsoft in late 2015, Eric Horvitz raised these concerns within the computer science community. In a piece he coauthored in an academic journal, he acknowledged that most computer scientists regarded the apocalyptic risks of a singularity as remote at best, but he said that the time had come to look more seriously at a growing range of other issues.[16] Satya picked up the mantle the next year, writing a piece in *Slate* suggesting that "the debate should be about the values instilled in the people and institutions creating this technology."[17] He offered some initial values that needed to be incorporated, including privacy, transparency, and accountability.

By late 2017, we concluded that what we really were talking about was the need for a full-blown approach to ethics across AI. It was far from a simple proposition. As computers gained the ability to make decisions previously reserved for humans, virtually every ethical question for humanity was becoming an ethical question for computing. If millennia of debate among philosophers had not forged clear-cut and universal answers, then a consensus was not likely to emerge overnight simply because we needed to apply them to computers.

By 2018, companies like Microsoft and Google that were at the forefront of AI started to address this new challenge head-on. Together with experts across academia and elsewhere, we recognized that we needed a set of ethical principles to guide our AI development. At Microsoft, we ultimately settled on six ethical principles for the field.

The first principle called on us to address the need for *fairness*, meaning the bias problem. We then moved to two other areas where there was already at least some public consensus—the importance of *reliability and safety* and the need for strong *privacy and security*. In important ways, these concepts had advanced through law and regulation in response to prior technological revolutions. Product liability and similar laws had set reliability and safety standards in response to the railroad and the automobile. Similarly, privacy and security norms had emerged in response to the communications and information technology revolutions. While AI was posing new challenges in these areas, we could build on these prior legal concepts.

Our fourth principle dealt with an issue that our employees had rallied behind since Satya became CEO in 2014. This was the importance of building *inclusive* technology that addressed the needs of people with disabilities. The company's focus on inclusive technology would naturally include AI. After all, if computers can see, imagine what that can do for people who are blind. And if computers can hear, imagine what that can mean for people who are deaf. And we didn't necessarily need to invent and distribute entirely new devices to pursue these opportunities. People were already walking around with smartphones that had cameras that could see and microphones that could hear. With inclusion as a fourth ethical principle, the path for progress in this area was already emerging.

While each of these four principles was important, we realized that they all rested on two other principles that were foundational to the others' success. The first was *transparency*. To us, this meant ensuring that information about how AI systems were making consequential decisions was public and understandable. After all, how can the public have confidence in AI and how can future regulators evaluate its adherence to the first four principles if the inner workings of AI are kept in a black box?

Some argue that AI developers should publish the algorithms they use, but our own conclusion was that these were likely to be less than enlightening in most circumstances or reveal valuable trade secrets that would undermine competition in the tech sector. We had already been working in the Partnership on AI with academics and other tech companies to develop better approaches. The emerging focus is on making AI explainable, like describing the key elements that are being used to make decisions.

The final ethical principle of AI would be the bedrock for everything else: *accountability.* Will the world create a future in which computers remain accountable to people and in which the people who design these machines remain accountable to everyone else? This may be one of the defining questions of our generation.

This final principle requires that humans remain in the loop so that AI-based systems can't go rogue without human review, judgment, and intervention. In other words, AI-based decisions that impact people's rights in a significant way need to remain subject to meaningful human review and control. And this requires people who are trained to evaluate the decisions that AI is making.

This also means, we concluded, that broader governance processes are critical. Every institution developing or using AI needs new policies, processes, training programs, compliance systems, and people to review and provide advice regarding the development and deployment of AI systems.

We published our principles in January 2018 and quickly realized that we'd struck a chord.[18] Customers asked for briefings about not only our AI technology but also our approach to ethical issues and practices. This made good sense. The company's whole strategy was to "democratize AI" by providing access to the technology's building blocks—tools for vision and speech recognition and machine learning, for example—so customers could create their own customized AI services. But this meant that a sophisticated approach to AI ethics needed to be developed and shared as broadly as the technology itself.

This broad dissemination of AI also meant that some regulation of the technology was not only likely, but essential. A general understanding about AI ethics could help encourage ethical people to act in an ethical way. But what about people who weren't interested in taking the high road? The only way to ensure that all AI systems worked in accordance with certain ethical standards was to require that they do so. And this meant putting the force of law and regulation behind the ethical standards that societies embrace.

It became apparent that regulatory efforts would proceed at a faster pace than the five-year time frame that I had predicted at Davos the preceding year. This was brought home in April 2018, when we were in Singapore and met government officials responsible for AI issues. "These issues can't wait. We need to get ahead of the technology," they told us. "We want to publish our first proposals not in a matter of years, but in a matter of months."

Inevitably, the ethical issues for AI will evolve from a general conversation to more concrete topics. In all probability, these will involve specific controversies. While it's impossible to predict exactly what we will all be debating five or ten years from now, we can draw insights from the issues that are emerging already.

In 2018, one of the first such controversies involved the looming use of AI in military weapons. As popularized in public debates, the issue concerned "killer robots," a phrase that conjures images from science fiction. It's a depiction that people readily can understand, perhaps reflecting the fact that the *Terminator* movie franchise has produced not only five sequels, but at least one film each decade since the original appeared in 1984. In other words, if you're a teenager or older, there is a good chance you've witnessed the dangers of autonomous weapons on the big screen.

One early lesson from this public policy debate is the need to develop a more nuanced understanding or classification of the types of technology involved. When I've talked with military leaders around the world, they all share one thing in common: No one wants to wake up in the morning to discover that machines have started a war while they were sleeping. Decision-making about war and peace needs to be reserved for humans.

But that doesn't mean that the world's military officials agree on everything else. This is where distinctions enter the picture. Paul Scharre, a former US defense official working at a think tank, brings to life increasingly pertinent questions in his book, *Army of None: Autonomous Weapons and the Future of War.*[19] As he illustrates, a central question is not just when but how computers should be empowered to launch a weapon without additional human review. On the one hand, even though a drone with computer vision and facial recognition might exceed human accuracy in identifying a terrorist on the ground, this doesn't mean that military officials need to or should take personnel and common sense out of the loop. On the other hand, if dozens of missiles are launched at a naval flotilla, the Aegis combat system's antimissile defenses need to respond according to computer-based decision-making. But even then the scenarios are varied and the use of the weapons system is customizable.[20] A human being should typically make the initial launch decision, but there isn't time for humans to approve each individual target.

Given the potential concerns about autonomous weapons, some have argued that tech companies should refuse all work with the military on AI-based technology. Google, for example, withdrew from an AI contract with the Pentagon after its employees protested.[21] We faced the same issue at Microsoft when some of our own employees raised similar concerns. We have long done work for the US and other militaries, a fact underscored when I toured the aircraft carrier USS *Nimitz* a few years ago in its home port in Everett, Washington, just north of Seattle. The carrier had more than four thousand computers running our Windows server operating system, powering a wide variety of the ship's functions.

But to many people, AI-based systems understandably fall into a different category from this type of platform technology. We recognized that new technology raised a new generation of complicated issues, and as we considered a potential contract to provide augmented reality technology and our HoloLens devices to soldiers in the US Army, we talked through what we should do.

In contrast to Google, we concluded that it was important for us to continue to provide our best technology to the US military as well as to other allied governments where we are confident in democratic processes and fundamental sensitivities around human rights. American and NATO military defenses have long depended on access to cutting-edge technology. As we stated both privately and publicly, "We believe in the strong defense of the United States and we want the people who defend it to have access to the nation's best technology, including from Microsoft."[22]

At the same time, we recognized that some of our own employees were uncomfortable working on defense contracts for the US or other military organizations. Some were citizens of other countries, some had different ethical views or were pacifists, and some simply wanted to devote their energy to alternative applications for technology. We respected these views, and we were quick to say that we would work to enable such individuals to work on other projects. Given Microsoft's size and diverse technology portfolio, we felt that we could most likely accommodate these requests.

But we also felt that none of this absolved us from the need to think through and engage on the complex ethical issues raised when one combines artificial intelligence with weapons. As we discussed this aspect within our senior leadership ranks, I pointed out that ethical issues had been important for weapons development since the 1800s, when hollowed bullets and then dynamite showed up on the battlefield. Satya reminded me that in fact ethical issues around war went back to the Roman writings of Cicero. He then sent me an email later that evening saying that his mother would not have been happy that he had remembered Cicero but forgotten the Indian Hindu epic, *The Mahabarta*. (Thankfully, he included a link to the entry in Wikipedia so I could learn more.[23])

These types of discussions led us to conclude that we needed to stay engaged on the ethical issues as active corporate citizens. And we believed that our presence could help shape the emerging public policy issues.[24] As we told our employees, we felt that no tech company had been more active in addressing the policy issues raised by new technology, especially for government surveillance and cyber weapons.[25] We felt that the best approach similarly was to advocate for responsible policies and laws regarding AI and the military.

We rolled up our sleeves to learn more and develop more refined views. This led us back to our six ethical principles that informed the ethical issues applicable to AI and weapons. We concluded that three were most at stake—reliability and safety, transparency, and most importantly, accountability. Only by addressing all three of these could anyone maintain public confidence that AI would be deployed in a way that would keep human beings in control.

We also found that there are some important similarities to the issues we have addressed in the context of security and nation-state cyberattacks. In that realm, there exist several domestic and international rules that already apply to new forms of technology, such as lethal autonomous weapons.

Many other dynamics also seem familiar to the security issues involving cyberweapons. UN Secretary-General António Guterres did not mince words in 2018 when he called for a ban on "killer robots," saying, "Let's call it as it is: The prospect of machines with the discretion and power to take human life is morally repugnant."[26] But, as is the case of cyberweapons, the world's leading military powers have resisted new international rules that would limit their technology development.[27]

This is leading to a discussion of the specific scenarios that go to the heart of potential concern—in the hope that this might break the logjam. Human Rights Watch, for example, has called for governments to "prohibit weapons systems that can select and engage targets without meaningful human control."[28] While there will likely be additional nuances to address, this type of international advocacy, with its focus on specific terms such as "meaningful human control," represents an essential element of what the world will likely need to address in this new generation of ethical challenges.

It's important that this work builds on existing ethical and human rights traditions. I've been impressed by the US military's deep and long-standing focus on ethical decision-making. This hasn't freed the armed forces from ethical lapses or, at times, enormous mistakes, but, as I learned from individuals ranging from senior generals to a West Point cadet, one can't graduate from an American military academy without taking a course in ethics.[29] The same is not yet true for computer science majors at many American universities.

As we've discussed these and similar emerging issues with leaders in other countries, we've come to appreciate that ethical views ultimately rest on broader human rights and philosophical foundations. This makes it essential to connect these topics to an understanding of the world's diverse cultures, as well as the varying laws and regulatory approaches this diversity creates.

Artificial intelligence, like all information technology, is designed to be global in character. The technologists who create it want it to work the same way everywhere. But laws and regulations can diverge among countries, leading to challenges for governmental diplomats and technologists alike. We had experienced this divergence on a recurring basis, first with intellectual property laws, then with competition rules, and most recently with privacy regulations. But in some respects, these differences are simple compared to the potential complexity of laws that address ethical issues that fundamentally are grounded in philosophy.

Now, as no technology before, artificial intelligence forces the world to confront the similarities and differences between these and other philosophical traditions. The issues raised by AI touch on topics like the role of personal accountability, the importance of public transparency, concepts of individual privacy, and notions of fundamental fairness. How can the world converge on a singular approach to ethics for computers when it cannot agree on philosophical issues for people? This is a fundamental conundrum for the future.

More than in the past this will require that those who create technology come not only from disciplines such as computer and data science but also from the social and natural sciences and humanities. If we're to ensure that artificial intelligence makes decisions based on the best that humanity has to offer, its development must result from a multidisciplinary process. And as we think about the future of higher education, we'll need to make certain that every computer and data scientist is exposed to the liberal arts, just as everyone who majors in the liberal arts will need a dose of computer and data science.

We'll also need to see more focus on ethics in computer and data science courses themselves. This might take the form of a focused course, or it might become an element in almost every course. Or both.

We can be optimistic that a new generation of students will embrace this cause with enthusiasm. In early 2018, I posed publicly a question together with Harry Shum, a Microsoft executive vice president responsible for much of our AI work, who earned his PhD in robotics. "Could we see a Hippocratic oath for coders like we have for doctors?" We joined others in suggesting that such an oath could make sense.[30] Within a matter of weeks, a computer science professor at the University of Washington had taken a stab at editing the traditional Hippocratic oath to suggest a new code for those who create artificial intelligence.[31] As Harry and I each spoke on college campuses around the world, we found that it was a question the next generation cared about.

Ultimately, a global conversation about ethical principles for artificial intelligence will require an even bigger tent. There will need to be seats at the table not only for technologists, governments, NGOs, and educators, but for philosophers and representatives of the world's many religions.

The need for this global conversation brought us to one of the last places I'd expect to talk about technology: the Vatican.

The visit had its share of ironies. We stopped in Rome in February 2019 just days before heading to Germany for the annual Munich Security Conference, where we'd be surrounded by the world's military leaders. And we were there to talk about ethics for computers with the Vatican's leaders just a week before they would host their own meeting to grapple with ethical issues for priests and the abuse of children in churches. The timing underscored humanity's aspirations and challenges.

When we pulled up to the Vatican, we were met by a beaming Monsignor Vincenzo Paglia, a snowy-headed, jovial archbishop of the Italian Catholic Church. The author of numerous books, he leads the Vatican's work addressing a variety of ethical issues, including new challenges relating to artificial intelligence. Microsoft and the Vatican had decided to cosponsor a doctoral dissertation award to explore the intersection between this emerging technology and long-standing ethical questions.

The afternoon provided a powerful reminder of the historical collisions between technology and science on the one hand and philosophy and religion on the other. While at the Vatican, Monsignor Paglia walked us through the Sistine library, where we carefully turned the pages of one of the first Bibles that Johannes Gutenberg had produced with the movable metal type of the printing press he invented and started using in the 1450s. It was a technological advance that revolutionized communications and impacted every part of European society, including the Church.

We then turned to a volume of letters written 150 years later. It preserved Galileo's correspondence with the Pope that was central to the dispute between Galileo and the Church regarding the respective place of the earth and the sun in the heavens. As the volume showed, Galileo had used his telescope in the early 1600s to document the changing position of sunspots, which showed that the sun was rotating. It was part of a bitter dispute about the interpretation of the Bible that led to Galileo's inquisition in Rome and his imprisonment under house arrest until he died.

The two books together illustrate how science and technology can connect or collide with matters of faith, religion, and philosophy. As with inventions like the printing press and the telescope, it's impossible to imagine that AI will leave these fields untouched. The question is how to promote a thoughtful, respectful, and inclusive global conversation.

It was a topic we discussed in a meeting with Pope Francis and Monsignor Paglia. We talked about the development of technology against the backdrop of nations increasingly turning inward, sometimes turning their backs on their neighbors and others in need. I mentioned Albert Einstein's dire warnings about the dangers of technology in the 1930s. The Pope then reminded me of what Einstein had said after the Second World War: "I know not with what weapons

World War III will be fought, but World War IV will be fought with sticks and stones."[32] Einstein's point was that technology, specifically nuclear technology, had progressed to the point where it could annihilate everything else.

As we left the meeting, Pope Francis shook my hand with his right hand and held my wrist with his left. "Keep your humanity," he urged.

As we think about the future of artificial intelligence, it's good advice for all of us.

Chapter 12　**AI AND FACIAL RECOGNITION: Do Our Faces Deserve the Same Protection as Our Phones?**

In June 2002, Steven Spielberg premiered a new movie he had directed, *Minority Report*, based on a famous 1956 short story by the science fiction writer Philip K. Dick. Set in 2054 in a crime-free Washington, DC, the film stars Tom Cruise, who plays the head of Precrime, an elite police unit that arrests killers before they commit their crimes. The team has the authority to make its arrests based on the visions of three clairvoyant individuals who can see into the future. But soon Cruise is evading his own unit—in a city where everyone and everything is tracked—when the psychics predict he will commit a murder of his own.[1]

More than fifteen years later, this approach to law enforcement happily seems far-fetched. But today, one aspect of *Minority Report* seems to be on track to arrive much earlier than 2054. As Cruise is on the run, he walks into the Gap. The retailer has technology that recognizes each entering customer and immediately starts displaying on a kiosk the images of clothes it believes the customer will like. Some people might find the offers attractive. Others might find them annoying or even creepy. In short, entering a store becomes a bit like we sometimes feel after browsing the web and then turning to our social media feed only to find new ads promoting what we just viewed.

In *Minority Report*, Spielberg asked theatergoers to think about how technology could be both used and abused—to eliminate crimes before they could be committed but also to abuse people's rights when things go wrong. The technology that recognizes Cruise in the Gap store is informed by a chip embedded inside him. But the real-world technology advances of the first two decades of the twenty-first century have outpaced even Spielberg's imagination, as today no such chip is needed. Facial-recognition technology, utilizing AI-based computer vision with cameras and data in the cloud, can identify the faces of customers as they walk into a store based on their visit last week—or an hour ago. It is creating one of the first opportunities for the tech sector and governments to address ethical and human rights issues for artificial intelligence in a focused and concrete way, by deciding how facial recognition should be regulated.

What started for most people as a simple scenario, such as cataloging and searching photos, has rapidly become much more sophisticated. Already many people have become comfortable relying on facial recognition rather than a password to unlock an iPhone or a Windows laptop. And it's not stopping there.

A computer can now accomplish what almost all of us as human beings have done almost since birth—recognize people's faces. For most of us, this probably began with the ability to recognize our mother. One of the joys of parenting comes when a toddler erupts enthusiastically when you return home. This reaction, which lasts until the onset of the teenage years, relies on the innate facial-recognition capabilities of human beings. While this is fundamental to our daily lives, we almost never pause to think about what makes it possible.

As it turns out, our faces are as unique as our fingerprints. Our facial characteristics include the distance of our pupils from each other, the size of our nose, the shape of our smile, and the cut of our jaw. When computers use photographs to chart these features and knit them together, they create the foundation for a mathematical equation that can be accessed by algorithms.

People are putting this technology to work around the world in ways that will make life better. In some cases, it may be a matter of consumer convenience. National Australia Bank, using Microsoft's facial-recognition technology, is developing the capability for you to walk up to an automated teller machine, or ATM, so you can withdraw money securely without a bank card. The ATM will recognize your face and you can then enter your PIN and complete your transaction.[2]

In other scenarios, the benefits are more far-reaching. In Washington, DC, the National Human Genome Research Institute is using facial recognition to help physicians diagnose a disease known as DiGeorge syndrome, or 22q11.2 deletion syndrome. It's a disease that more often afflicts people who are African, Asian, or Latin American. It can lead to a variety of severe health problems, including damage to the heart and kidneys. But it also often manifests itself in subtle facial characteristics that can be identified by computers using facial-recognition systems, which can help a doctor diagnose a patient in need.[3]

These scenarios illustrate important and concrete ways that facial recognition can be used to benefit society. It's a new tool for the twenty-first century.

Like so many other tools, however, it can also be turned into a weapon. A government might use facial recognition to identify every individual attending a peaceful rally, following up in ways that could chill free expression and the ability to assemble. And even in a democratic society, the police might rely excessively on this tool to identify a suspect without appreciating that facial recognition, like every technology, doesn't always work perfectly.

For all these reasons, facial recognition easily becomes intertwined with broader political and social issues and raises a vital question: What role do we want this form of artificial intelligence to play in our society?

A glimpse of what lies ahead emerged suddenly in the summer of 2018, in relation to one of the hottest political topics of the season. In June, a gentleman in Virginia, a self-described "free software tinkerer," also clearly had a strong interest in broader political issues. He had posted a series of tweets about a contract Microsoft had with the US Immigration and Customs Enforcement, or ICE, based on a story posted on the company's marketing blog in January.[4] It was a post that frankly everyone at the company had forgotten. But it says that Microsoft's technology for ICE passed a high security threshold and will be deployed by the agency. It says the company is proud to support the agency's work, and it includes a sentence about the resulting potential for ICE to use facial recognition.[5]

In June 2018, the Trump administration's decision to separate children from parents at the southern US border had become an explosive issue. A marketing statement made several months earlier now looked a good deal different. And the use of facial-recognition technology looked different as well. People worried about how ICE and other immigration authorities might put something like facial recognition to work. Did this mean that cameras connected to the cloud could be used to identify immigrants as they walked down a city street? Did it mean, given the state of this technology, with its risk of bias, that it might misidentify individuals and lead to the detention of the wrong people? These were but two of many questions.

By dinnertime in Seattle, the tweets about the marketing blog were tearing through the internet, and our communications team was working on a response. Some employees on the engineering and marketing teams suggested that we should just pull the post down, saying, "It is quite old and not of any business impact at this point."

Three times, Frank Shaw, Microsoft's communications head, advised them not to take it down. "It will only make things worse," he said. Nonetheless, someone couldn't resist the temptation and deleted part of the post. Sure enough, things then got worse and another round of negative coverage followed. By the next morning, people had learned the obvious lesson and the post was back up in its original form.

As so often happens, we had to sort out what the company's contract with ICE really covered.

As we dug to the bottom of the matter, we learned that the contract wasn't being used for facial recognition at all. Nor, thank goodness, was Microsoft working on any projects to separate children from their families at the border. The contract instead was helping ICE move its email, calendar, messaging, and document management work to the cloud. It was similar to projects we were working on with customers, including other government agencies, in the United States and around the world.

Nonetheless, a new controversy was born.

Some suggested that Microsoft cancel our contract and cease all work with ICE, a persistent theme about government use of technology that would take hold that summer. One group of employees circulated a petition to halt the ICE contract. The issue began to roil the tech sector more broadly. There was similar employee activism at the cloud-based software company Salesforce, focused on its contract with US Customs and Border Protection. This followed employee activism at Google, which had led the company to cancel a project to develop artificial intelligence for the US military. And the ACLU targeted Amazon, backing Amazon employees who voiced concern about Rekognition, its facial-recognition service.[6]

For the tech sector and the business community more broadly, this type of employee activism was new. Some saw a connection to the role that unions had played in certain industries for well over a century. But unions had focused principally on the economic and working conditions of their members. Employee activism in the summer of 2018 was different. This activism called on employers to adopt positions on specific societal issues. The employees had nothing directly or even indirectly to gain. They instead wanted their employers to stand up for societal values and positions that they thought were important.

It was helpful for us to take stock of the different reactions to this new wave of employee activism. Just a few miles away in Seattle, the leaders at Amazon seemed to do less to engage directly with employees to discuss these types of issues.[7] That reaction appeared to dampen some of the employee interest in raising issues, in effect encouraging people to keep their heads down and focused on business. In Silicon Valley, the leaders at Google took a very different approach, sometimes responding quickly to employee complaints by reversing course, including by pulling the plug on an AI-focused military contract.[8] It was quickly apparent that there was no single approach, and every company needed to think about its own culture and what it wanted in terms of its connection to its employees. As we thought about our own culture, we decided to chart a path between the approaches we were watching elsewhere.

These episodes seemed to reflect several important developments. First and perhaps most important was the rising expectation that employees had for their employers. This had been captured well a few months earlier when the annual Edelman Trust Barometer identified the change.[9] The Edelman communications firm has been publishing its Trust Barometer since 2001, identifying changes in the public mood around the world as people's trust in institutions waxes and wanes. Its report in early 2018 showed that while trust in many institutions had plummeted, employee confidence in employers was a big outlier. It found that worldwide 72 percent of people trusted their employer "to do what is right," with an even higher 79 percent feeling that way in the United States.[10] In contrast, only a third of Americans felt that way about their government.

What we were experiencing reflected this view and went even further. In the tech sector, some employees wanted to play an active role in shaping their companies' decisions and engagement on the issues of the day. Perhaps not surprisingly, this view was more pronounced at a time when people had less trust in governments. Employees were looking to another institution they hoped might do the right thing and have some influence on public outcomes.

The change thrust business leaders into new terrain. At a small dinner I attended in Seattle, the CEO of one tech company summed up the collective angst. "I feel well prepared for most of my job," he said, describing how he'd risen up the ranks. "But now I'm being thrust into something completely different. I really don't know how to respond to employees who want me to take on their concerns about immigration, climate issues, and so many other problems."

Perhaps not surprisingly, the phenomenon was also the most pronounced with our newest generation of employees. After all, there's a well-established tradition of students clamoring for societal change on college campuses, at times pushing their universities to lead the way by changing their policies. Because it was summer, we had roughly three thousand interns working on the Microsoft campus. Not surprisingly, they took a strong interest in the issue. Some wanted to have a direct impact on the company's position even if they were just spending the summer with us.

We talked about how to think through the topic and respond. As Satya and I compared notes, I reflected on what I had learned serving on Princeton University's board of trustees. "I think leading a tech company is becoming more like leading a university," I said. "We have researchers with PhDs who are like the faculty. We have interns and young employees who sometimes have views similar to university students. Everyone wants to be heard, and some want us to boycott a government agency much like they want a university to boycott the purchase of stock in a company that's doing something objectionable."

For me, there had been a couple of key takeaways from my trustee experience. Perhaps the most important was that well-intentioned students might not have all the right answers, but they might be asking the right questions. And these questions could lead to a better path that had eluded experts and senior leaders alike. As I like to say to our teams within the company, the best response to a half-baked idea often is not to kill the idea, but to finish baking it. Some of our best initiatives came together this way. And it built on the culture that Satya had fostered for Microsoft, grounded in a growth mind-set and constant learning. In short, if a new era of employee activism was dawning, it would be important for us to find new ways to engage with our employees, understand their concerns, and try to develop a thoughtful answer.

I had also learned from my Princeton experience that universities had developed some sound processes to meet this need. They created opportunities for everyone to have input and have more collaborative discussion. It allowed emotions to subside and encouraged reason to prevail by helping a group think through and make a difficult decision with the time needed to get it right. We set out on this path, and Eric Horvitz, Frank Shaw, and Rich Sauer, our senior lawyer responsible for AI ethics issues, started holding a series of roundtables that employees could attend.

It became increasingly important to spell out when we thought it made sense for the company to take a position on a public issue and when we should not. We didn't view corporate leadership as a license to use the company's name to address any issue under the sun. There needed to be some vital connection to us. We felt our responsibility was fundamentally to address public issues that impacted our customers and their use of our technology, our employees both at work and in their community, and our business and the needs of our shareholders and partners. This didn't answer every question, but it provided a useful framework for discussions with our employees.

Employee questions also pushed us in a constructive way to think harder about our relationship with the government and the challenges posed by new technology such as facial recognition.

On the one hand, we were not comfortable with the suggestion that we react to the events of the day by boycotting government agencies, especially in democratic societies governed by the rule of law. In part this was a principled reaction. As I often tried to remind people, no one elected us. It seemed not just odd but undemocratic to want tech companies to police the government. As a general principle, it seemed more sensible to ask an elected government to regulate companies than to ask unelected companies to regulate such a government. Satya and I discussed this point frequently and we believed it was important.

There was a pragmatic aspect as well. We recognized the enormous dependence that organizations and individuals had on our technology. It was far too easy to unleash chaos and unintended consequences if we simply turned technology off based on an objection to something a government agency was doing.

This pragmatic dimension was thrust into bold relief in August 2018. As I drove to work on a Friday morning, I listened to an account on *The Daily* podcast from the *New York Times* that got to the heart of the matter. The issue of the day was the government's inability to meet a court deadline to reunite immigrant children with their families. As I listened, I recognized the voice of Wendy Young, who leads Kids in Need of Defense, or KIND, a pro bono organization I have chaired for more than a decade.[11] As Wendy explained, the administration had implemented the initial family separation policy "with no thought given to how you reunify families" later.[12]

While I was familiar with this situation based on several conversations with Wendy, I was struck by an additional detail reported by *New York Times* journalists Caitlin Dickerson and Annie Correal. They explained that the Customs and Border Protection personnel used a computer system with a drop-down menu when people initially crossed the border. Agents would classify someone either as an unaccompanied minor, an individual adult, or an adult with children, meaning a family unit. When children subsequently were separated from their parents, the computer system's design forced agents to go back and change this designation, for example by inputting a child's name as an unaccompanied minor and the parent's name as an individual adult. Critically, this overwrote the prior data, meaning the system no longer retained the family designation that previously had listed everyone together. As a result, the government no longer had any record that connected family members.

This was not only a story about immigration and families. It was also a story about technology. The government was using a structured database that worked for one process but not for another. Rather than update the IT system to support the new steps involved in separating families, the administration had plunged ahead without thinking about the computer architecture that would be needed. Having seen CBP's systems at a command center near the Mexican border on a visit with Wendy just months before, I was not surprised that its systems were antiquated. But I was still horrified that the administration had failed to think about the implications of what it needed in terms of basic technology infrastructure.

When I walked into the conference room that morning where Satya's senior leadership team was gathering for our Friday meeting, I shared what I had heard. As we talked about it, we recognized that it connected to our broader concerns about the proposition advocated by some that tech companies take it upon themselves to unplug government agencies from all services based on policies to which we object. Technology has become a key infrastructure of our lives, and the failure to update it—or worse, a decision simply to unplug it—could have all kinds of unintended and unforeseen consequences. As Satya had noted several times in our internal conversations, the government was using email as one tool to bring families back together. If we shut it off, who knew what would happen?

This led us to conclude that boycotting a government agency in the United States was the wrong approach. But the people advocating for such action, including some of our own employees, were asking some of the right questions. Facial-recognition technology, for example, created challenges that needed more attention.

As we thought it through, we concluded that this new technology should be governed by new laws and regulations. It's the only way to protect the public's need for privacy and address risks of bias and discrimination while enabling innovation to continue.

To many, it was odd for a company to call on the government to regulate its products. John Thompson, our board chair, said that some people in Silicon Valley told him that they assumed we were behind other companies in the market and wanted regulation to slow our competitors down. This made me bristle. To the contrary, in 2018 the National Institute of Standards and Technology completed another round of facial-recognition testing, finding that our algorithms were at or near the top in every category.[13] While forty-four other companies had provided their technology for testing, many others, including Amazon, had not.

Our interest in regulation came from our emerging sense of where the market was heading. A few months earlier, one of our sales teams had wanted to sell an AI solution that included facial-recognition services to the government of a country that lacked an independent judiciary and had a less than stellar track record for respecting human rights. The government wanted to deploy the service with cameras across its capital city. Our concern was that a government that flouted human rights could use the technology to follow anyone anywhere—or everyone everywhere.

With the advice of our internal AI ethics committee, we decided we would not move forward with the proposed deal. The committee had recommended that we draw a line and refrain from making facial-recognition services available for generalized use in countries that Freedom House, an independent watchdog that tracks freedom and democracy around the world, had concluded were not free. The local team was not happy. As the person responsible for the final call, I received an impassioned email from the head of the sales team that had been working on the deal. She wrote that "as a mother and a professional," she "would have felt much safer" if we had made the service available to counter risks of violence and acts of terror.

I understood her point. It underscored the difficult trade-offs that had characterized the age-old tensions between public safety and human rights. It also illustrated the subjective nature of many of the new ethical decisions that will be made for artificial intelligence. And, of course, we remained concerned that, as she and others had pointed out, if we refused to provide this service, some other company might step in. In that case, we would both lose the business and then watch on the sidelines as someone else facilitated the harmful use despite our position. But as we had balanced all these factors, we concluded that we needed to try to nudge the development of this new technology toward some type of ethical foundation. And the only way to do this was to turn down certain uses and push for a broader public discussion.

This need for a principled approach was reinforced when a local police force in California contacted us and said they wanted to equip all their cars and body cameras with a capability to take a photo of someone pulled over, even routinely, to see if there was a match against a database of suspects for other crimes. We understood the logic but advised that facial-recognition technology remained too immature to deploy in this type of scenario. Use of this nature, at least in 2018, would result in too many false positives and flag people who had been wrongly identified, especially if they were people of color or women, for whom there remained higher error rates. We turned down the deal and persuaded the police force to forgo facial recognition for this purpose.

These experiences started to provide some insights into principles we could apply to facial recognition. But we worried that there would be little practical impact if we took the high road only to be undercut by companies that imposed no safeguards or restrictions at all, whether those companies were on the other side of Seattle or on the other side of the Pacific. Facial recognition, like so many AI-based technologies, improves with larger quantities of data. This creates an incentive to do as many early deals as possible and hence the risk of a commercial race to the bottom, with tech companies forced to choose between social responsibility and market success.

The only way to protect against this race to the bottom is to build a floor of responsibility that supports healthy market competition. And a solid floor requires that we ensure that this technology, and the organizations that develop and use it, are governed by the rule of law.

We drew insights from the historical regulation of other technologies. There are many markets in which a balanced approach to regulation has created a healthier dynamic for consumers and producers alike. The auto industry spent decades in the twentieth century resisting calls for regulation, but today there is broad appreciation of the essential role that laws have played in ensuring ubiquitous seat belts and air bags and greater fuel efficiency. The same is true for air safety, food, and pharmaceuticals.

Of course, it was one thing to talk about the need for regulation and another to define what type of regulation would be most sensible. In July 2018, we published a list of questions that we thought needed to be considered [14] and asked people for advice about possible answers. The discussions started with employees and technology experts, but quickly expanded across the country and around the world, including civil liberties groups like the ACLU, which was playing an active role on the issue.

I was particularly struck by the reaction of legislators I met with in the National Assembly in Paris. As one member said, "No other tech company is asking us these questions. Why are you different?" Facial recognition was the type of issue where we sometimes diverged from others in the tech sector. Perhaps more than anything else, this reflected what we had learned from our antitrust battles in the 1990s. At that time, we had argued, like many companies and industries, that regulation was unnecessary and likely to be harmful. But one of the many lessons we'd learned from that experience was that such an approach didn't necessarily work—or would be regarded as unacceptable—for products that have a sweeping impact across society or that combine beneficial and potentially troubling uses.

We no longer shared the resistance that most tech companies traditionally had shown for government intervention. We'd already fought that battle. Instead we had endorsed what we thought of as a more active but balanced approach to regulation. That was one reason we called for federal privacy legislation in the United States as early as 2005. We knew there would be days when the government would get the details wrong and when we might regret advocating its involvement. But we believed this general approach would be better for technology and society than a practice that relied exclusively on the tech sector to sort everything out by itself.

The key was to figure out the specifics. A piece by Nitasha Tiku in *Wired* captured the importance of this dynamic. As she noted toward the end of 2018, "After a hellish year of tech scandals, even government-averse executives have started professing their openness to legislation."[15] But, as she recognized, our goal was to take "it one step further" by proposing a specific proposal for governments to regulate facial-recognition technology.

By December we felt we had learned enough to suggest new legislation. We knew we didn't have answers for every potential question, but we believed there were enough answers for good initial legislation in this area that would enable the technology to continue to advance while protecting the public interest. We thought it was important for governments to keep pace with this technology, and an incremental approach would enable faster and better learning across the public sector.

In essence, we borrowed from a concept that has been championed for start-up companies and software development, referred to as a "minimum viable product." As defined by entrepreneur and author Eric Ries, it advocates creating "an early version of a new product that allows a team to collect the maximum amount of validated learning (learning based on real data gathering rather than guesses about the future) about customers."[16] In other words, don't wait until you have the perfect answer to every conceivable question. If you are confident that you have reliable answers to critical questions, act on them, build your product, and get it into the market so you can learn from real-world feedback. It's an approach that has enabled not just businesses but technology to move faster and more successfully.

Even while moving more quickly, it's critical to be thoughtful and confident that the initial steps will be positive. In this case, we believed we had a strong set of ideas to address facial recognition. I publicly made our case for new legislation at the Brookings Institution in Washington, DC,[17] and published more details about our proposal.[18] We then took the cause on the road, presenting it over the next six months at public events and legislative hearings across the United States and in eight other countries around the world.

We believed that legislation could address three key issues—the risk of bias, privacy, and the protection of democratic freedoms. We believed that a well-functioning market could help accelerate progress to reduce bias. No customer we encountered was interested in buying a facial-recognition service that had high error rates and resulted in discrimination. But the market couldn't function if customers lacked information. Just as groups such as Consumer Reports had informed the public about issues like auto safety, we believed academic and other groups could test and provide information on the accuracy of competing facial-recognition services. This would further empower researchers like Joy Buolamwini at the Massachusetts Institute of Technology to pursue research that would prod us along. The key was to require companies that participated in the market to make it possible to test their products. That's what we proposed, in effect using regulation to reinforce the market.[19]

To help reduce the risk of discrimination, we believed a new law should also require organizations that deploy facial recognition to train employees to review results before making key decisions—rather than just turning decision-making over to computers.[20] Among other things, we were concerned that the risks of bias could be exacerbated when organizations deployed facial recognition in a manner that is different from what was intended when the technology was designed. Trained personnel could help address this problem.

In some ways, a thornier question was when law enforcement should be permitted to use facial recognition to engage in ongoing surveillance of specific individuals as they go about their day.

Democracy has always depended on the ability of people to meet and talk with each other and even to discuss their views both in private and in public. This relies on people being able to move freely and without constant government surveillance.

There are many governmental uses of facial-recognition technology that protect public safety and promote better services for the public without raising these types of concerns.[21] But when combined with ubiquitous cameras and massive computing power and storage in the cloud, facial-recognition technology could be used by a government to enable continuous surveillance of specific individuals. It could do this at any time or even all the time. This use of such technology in this way could unleash mass surveillance on an unprecedented scale.

As George Orwell described in his novel *1984*, one vision of the future would require citizens to evade government surveillance by finding their way secretly to a blackened room to tap in code on each other's arms—because otherwise cameras and microphones will capture and record their faces, voices, and every word. Orwell sketched that vision nearly seventy years ago. We worried that technology now makes that type of future possible.

The answer, in our view, was for legislation to permit law enforcement agencies to use facial recognition to engage in ongoing surveillance of specific individuals only when it obtains a court order such as a search warrant for this monitoring or when there is an emergency involving imminent danger to human life. This would create rules for facial-recognition services that are comparable to those now in place in the United States for the tracking of individuals through the GPS locations generated by their cell phones. As the Supreme Court had decided in 2018, the police cannot obtain without a search warrant the cell phone records that show the cell sites, and hence the physical locations, where someone has traveled.[22] As we put it, "Do our faces deserve the same protection as our phones? From our perspective, the answer is a resounding yes."[23]

Finally, it was apparent that the regulation of facial recognition should protect consumer privacy in the commercial context as well. We're rapidly entering an era in which every store can install cameras connected to the cloud with real-time facial-recognition services. From the moment you step into a shopping mall, it's possible not only to be photographed but to be recognized by a computer wherever you go. The owner of a shopping mall can share this information with every store. With this data, shop owners can learn when you visited them last and what you looked at or purchased, and by sharing this data with other stores, they can predict what you're looking to buy next.

Our point was not that new regulations should prohibit all such technology. To the contrary, we are among the companies working to help stores responsibly use technology to improve the shopping experience. We believe many consumers will welcome the resulting customer service. But we also felt that people deserve to know when facial recognition is being used, ask questions, and have real choices.[24]

We recommended that new laws require organizations that use facial recognition to provide "conspicuous notice" so people will know about it.[25] And we said there needed to be new rules developed to decide when and how people can exercise meaningful control and provide consent in such contexts. The latter issue clearly will require additional work over the coming years to define the right legal approach, especially in the United States where privacy laws are less developed than in Europe.

It was also helpful to think about the reach of new laws. For some aspects, we didn't need to encourage the passage of laws everywhere. For example, if one significant state or country were to require that companies make their facial-recognition services available for public and academic testing, then the results could be published and would spread everywhere else. Acting on this belief, we encouraged state legislators to consider new legislation as they prepared to convene for their sessions across the United States at the start of 2019.[26]

But when it comes to consumer privacy protection and the protection of democratic freedoms, one needs new laws in every jurisdiction. We recognized that this is likely unrealistic, given the differing views of governments around the world. For this reason, a simple call for the government to act would never be enough. Even if the US government got its act together, it's a big world. People could never have confidence that all the world's governments would use this technology in a way that is consistent with human rights protections.

The need for government leadership does not absolve technology companies of our own ethical responsibilities. Facial recognition should be developed and used in a manner consistent with broadly held societal values. We published six principles corresponding to our legislative proposals, which we have gone on to apply to our facial-recognition technology, and we have created systems and tools to implement them.[27] Other tech companies and advocacy groups have started to adopt similar approaches.

The facial-recognition issue provides a glimpse into the likely evolution of other ethical challenges for artificial intelligence. While one can start, as we did, with broad principles that are applicable across the board, these principles are tested when put into practice around concrete AI technologies and specific scenarios. That's also when potentially controversial AI uses are more likely to emerge.

There will be more issues. And as with facial recognition, each will require detailed work to sift through the potential ways the technology will be used. Many will require a combination of new regulation and proactive self-regulation by tech companies. And many will raise important and differing views between countries and cultures. We will need to develop a better capability for countries to move more quickly and collaboratively to address these issues on a recurring basis. That's the only way we'll ensure that machines remain accountable to people.

Chapter 13 ⋙ **AI AND THE WORKFORCE:**
The Day the Horse Lost Its Job

On December 20, 1922, stamping hooves echoed through the streets of Brooklyn Heights as fire engine 205's finest strained against their worn harnesses, anxious to charge into the cold winter morning. The assistant fire chief, "Smokey Joe" Martin, sounded the station's bell, signaling the company's first whip to belt out a "Haw!" and drive the eager team of fire horses into New York's streets.

But there was no fire to fight. The horse-drawn engine was headed to the Brooklyn Borough Hall, where it would hand the reins over to a waiting motorized coach.

As the horse-drawn steam engine and hose wagon clamored out of the firehouse, New Yorkers lined the sidewalks to cheer and chase the team as it hurtled through the city. Citizens, local officials, and firefighters, including the firehouse's beloved dalmatian Jiggs, had turned out in force to pay tribute to engine 205's last "faithful and true" fire horses.[1]

As the drenched, heaving team pulled up to the hall, Jiggs anxiously circled the fire engine, prodding the men to hook the hose up to a hydrant.[2] Instead, firemen draped the horses with wreaths of flowers. It would be the team's final call—and the final run for all fire horses in New York City.

While putting the fabled fire horse out to pasture was a practical matter, progress, as the *Brooklyn Eagle* wrote, had a profound impact on the city's culture. "To the small boys of three generations the fire horse has been a delight as the fireman has been an inspiration. Today the fire horse vanishes in New York City, probably forever."[3]

After more than fifty years of service, the fire horse had lost its job. It was a story about changing technology and its impact on work. Fire horses themselves had previously replaced men in pulling fire engines. Volunteer teams of men and boys had originally pulled these fire engines, but in 1832, when the New York fire department's force was depleted by the city's cholera epidemic, horsepower came to the rescue. "Not enough men . . . could be mustered to drag the engine to the scene of the conflagration." Necessity, being the mother of invention, forced the FDNY to spend a hefty sum of $864 to purchase a fleet of horses to replace the sick and dying firemen.[4]

It wasn't until the 1860s that manpower was officially swapped out for horsepower in firehouses. But the transition wasn't easy. One obstacle was the firefighters' pride in their work as haulers. In 1887, Abraham Purdy, known at the time as one of the oldest living firemen, said the introduction of horses had created so many squabbles within the fire department that members resigned.[5]

But nothing could stem the tide of progress. Continuing advancements in equipment, including quick-hitch horse collars, eventually allowed horses to relieve volunteers of their duties hauling hoses by hand. By 1869, well-trained horses and men could exit a firehouse in less than a minute.[6] In the twentieth century, however, horses met the same fate in pulling a fire wagon as people had the century before: They were replaced in their jobs. This time, the replacement was a machine powered by a combustion engine.

This represented a small slice of a large economic pie. The technological changes of almost three centuries have repeatedly altered the nature of work and indisputably raised standards of living overall. But the unavoidable truth is that there have always been winners and losers. Sometimes these winners or losers are individuals and families. Often, they are communities, states, and even countries.

Today, the world understandably eyes artificial intelligence with a similar mixture of hope and anxiety. Will computers do to us what machines did to horses? To what extent are our jobs at risk?

These are questions we find people asking us everywhere we travel. They were top of mind as our flight descended on a gusty Sunday afternoon toward the high-desert runway in El Paso, the West Texas town that sits on the border of Mexico. The plane bobbed in the crosswinds, bouncing hard as it landed. The city sits at a break in the craggy Franklin Mountains, serving as an intersection of sorts for two states and two nations along the Rio Grande River. We had taken in much of this broad vista through the plane's window.

Our inhospitable landing was quickly offset by the warm welcome we received. El Paso is a vibrant bilingual, bicultural city that is part of a singular international community that encompasses the larger city of Juarez on the Mexican side of the border.

We'd traveled to the region as part of Microsoft's TechSpark program, an initiative we created in 2017 to partner with a half dozen communities across the United States.[7] Our goal was to work in new ways with local business, government, and nonprofit leaders to better assess the impact technology had on communities outside the country's largest cities. It included an innovative tech partnership with the Green Bay Packers, near where I had grown up. Across the country, it created an opportunity to learn about the new challenges that tech was creating and the brighter prospects that might emerge if we could harness technology in new ways.[8]

As we drove the Interstate 10 corridor, a recent addition to the El Paso economy caught our eye. Massive call centers had sprung up in the desert, a rapidly growing industry drawing on the region's ability to recruit workers who speak both English and Spanish. These call centers, which employ thousands of El Pasoans, can serve a population of almost a billion people throughout the Western Hemisphere. But as we toured the area, a nagging thought weighed heavily, and it wasn't a happy one. Many of these call center jobs could vanish a decade from now, maybe even sooner, replaced by artificial intelligence.

As we met with the region's leaders to talk about how AI was likely to impact the local economy, it seemed important to start with a caveat. There is no crystal ball. If anything, it's too easy for a tech leader to lay claim as a great "futurist" and offer seemingly confident and even grandiose predictions about what the world will look like in a decade or two. People will undoubtedly listen, and the good news, if you pursue this approach, is that few are likely to remember a decade from now what you've said. And even if you're completely wrong, there's plenty of time for a course correction.

But despite this caveat, there is an important opportunity to draw real insights to help predict where the future is heading. As we met with people in El Paso and discussed what AI might mean for jobs in the region, we talked about two places we can look for insights.

The first is to understand what AI can and cannot do well and appreciate how this will impact jobs and work. To state the obvious, AI will most readily replace jobs that involve functions that it can perform well. It makes sense to consider the recent advances that have enabled AI to understand human speech, recognize images, translate languages, and reach new conclusions based on an ability to discern patterns. If a large part of a job involves tasks that can be completed by AI—and faster—then that job is probably at risk of being replaced by a computer.

If we had to predict a job that AI is likely to eliminate sooner than most, we'd nominate the position of taking orders from customers in the drive-through window at a fast-food restaurant. Today, a human being listens to what we say and enters our order into a computer. But with improved outdoor microphones, artificial intelligence can capture and understand the spoken word as well as a person, meaning that this task could soon be performed entirely by a machine. Before we know it, we'll approach a drive-through and talk to a computer rather than a person. The computer may not be 100 percent accurate, but neither is a human being. That's why there will be an opportunity to verify and correct our order.

It's why we looked at the burgeoning call-center industry in El Paso with a mixture of admiration and concern. So much of talking to customers on the phone involves understanding what they want and solving their problems. But computers are handling simple customer support requests already. Often it feels as if the hardest thing to do when calling for customer support is reaching a real person. That's because computers are answering the phone, asking us to enter a number as a command, and deciphering our spoken words in the form of simple sentences. As AI continues to improve, more of these tasks will be automated.

This also points to other categories of jobs that may be at risk. Much of driving involves recognizing images visually through the windows of a car, analyzing this information, and making decisions. As computers progress in these areas, AI can take over the function of driving a car or truck. In the middle of the twentieth century, it was common for people to get paid to sit in and operate elevators in tall buildings. Today, that seems not just quaint, but anachronistic. By the middle of the twenty-first century, will people feel the same way about a human taxi or Uber driver?

A similar phenomenon is already impacting the inspection of

machines. At Microsoft we have more than thirty-five hundred fire extinguishers on our campus in Redmond. We used to employ people to inspect each fire extinguisher's pressure every month to ensure it hadn't fallen below an acceptable threshold. Today they're all connected by small sensors into the company's network. Whenever the pressure falls below a certain level, a centralized dashboard flags it immediately so someone can fix it. Safety is enhanced and costs fall. But people are no longer employed to inspect those thirty-five hundred fire extinguishers each month.

While machines and automation have long been replacing jobs that involve rote tasks or repetitive manual labor, the ability of computers to think means that we'll see jobs at risk that involve brains as well as brawn. For example, given the rapidly improving ability of AI to translate human languages, it seems inevitable that the jobs of human interpreters will increasingly be at risk.

Similarly, take the work of a paralegal. Technology-based services have been impacting this area for years. Fifteen years ago, we employed almost one paralegal for every lawyer at Microsoft. But the ability to offer self-service-based assistance on an internal network has contributed to our needing today only one paralegal for every four lawyers. And with the growing ability of AI-based systems to master pattern recognition through machine learning, there is good reason to expect technology to continue to absorb tasks performed not only by paralegals but also by junior attorneys doing legal research.

Even an advanced degree or sophisticated set of skills won't keep workers from being robbed of their jobs. AI will impact every level of the income scale. Take radiologists, who today earn an average of four hundred thousand dollars a year.[9] They spend much of their day scanning CT and MRI scans looking for abnormalities. If you feed enough images into an AI-powered machine, it can be trained to identify normal and abnormal X-rays, whether they involve broken bones, hemorrhages, or cancerous tumors.[10]

At one level, AI's destruction of some jobs is clearly daunting, but there are some silver linings in this cloud. Having started out as a junior lawyer myself, I can see why many law school graduates might endorse the proposition that an awful lot of the early legal work in one's career is downright boring. I still recall my own reaction in 1986 when one of my first assignments in a large law firm involved reading and dictating summaries of more than a hundred thousand pages of documents—work that's already automated today. Stimulating moments most typically came not in chasing answers through volumes of documents or legal cases, but in creatively crafting the right questions to ask in the first place. In some instances, AI will replace the world's mundane activities and chores, enabling us to elevate our thinking and focus on tasks that are more edifying.

In many ways, human beings have been remarkably resilient in creating new tasks that require more time and attention. The arrival of cars, calculators, voice mail, word processing, and graphic design software may have eliminated and changed many jobs over the decades, but there still has been plenty of work to go around. As some have noted, jobs are a bundle of tasks. Some tasks can be automated, while others cannot.[11]

After so many waves of industrialization and automation, what is it that is so demanding of our time? As Rick Rashid, a former head of Microsoft Research, observed only half-jokingly several years ago, a lot more people now spend a lot more time in meetings. And it's not just meetings that occupy our time. We all focus much more energy on communicating with each other in plenty of other ways as well. In offices, the average worker receives and sends 122 business emails each day.[12] By 2018, the people of the planet were creating a whopping 281 billion business and consumer emails every day.[13] But that's just a slice of people's communications. On a daily basis, people around the world are also sending 145 billion SMS and in-app messages.[14]

In an important way, this is the other side of the coin. There are certain tasks that AI likely won't perform well. Many of these involve soft skills such as collaboration with other people, which will remain fundamental in organizations large and small. As Rick recognized, this often requires meetings (hopefully well-planned ones). AI similarly is unlikely to excel in providing the empathy required of nurses, counselors, teachers, and therapists. Each of these individuals will likely use AI for some tasks, but it seems unlikely that it could replace their work entirely.

Like all new technologies, AI will not just eliminate and change jobs, it will also create new industries and careers. But knowing what new jobs it will create is much harder than analyzing its potential impact on today's workforce. Still, new jobs involving AI itself are already starting to emerge.

We've stumbled upon some of these insights ourselves when discussing AI with political leaders around the world.

One such opportunity arose in the spring of 2017 when I visited Microsoft's subsidiary in the United Kingdom and we hosted a visit by Prime Minister Theresa May. As I stood next to Cindy Rose, the CEO of our UK business, we both held our breath a bit as we watched a young apprentice place a HoloLens headset on the prime minister's head. We exhaled when the prime minister moved briskly through an augmented-reality demonstration of how the device could be used to identify faults in sophisticated machinery. (As it turned out, the HoloLens was far easier to master than devising a negotiating strategy for Brexit.)

After the demo, Prime Minister May took off the headset and turned to our apprentice to ask him about his job. He replied proudly, "I'm an envisioning adviser. I help customers envision how they can take new technology like augmented reality and use it inside their company."

"An envisioning adviser," the prime minister repeated. "That's a job I've never heard of."

There will be many new jobs with new names that are unfamiliar to us today. Our friends—or our children's friends—will turn up at parties and describe their roles as facial-recognition specialists, augmented-reality-based architects, and IoT data analysts. As was the case for generations past, there will be days when we feel the need for an updated dictionary to understand what people are describing.

Ultimately everyone would like a precise prediction about these new jobs. But unfortunately, the future, like the past, is messy. No one is clairvoyant.

This point was brought home in the fall of 2016, when Satya and I met with German chancellor Angela Merkel in her office in the glass and polished steel chancellery in Berlin. The building had opened in 2001 near the much older Reichstag, a symbol of the German nation dating to the end of the nineteenth century.

With a portrait of Germany's famous postwar chancellor, Konrad Adenauer, gazing down on us, we were joined at the conference table by an interpreter whose strong command of both German and English, we quickly realized, was more than matched by her diplomatic professionalism. While Merkel spoke English well, and much better than our few words of German, some of the conversation was sufficiently technical that it benefited from an interpreter. At one point, Satya talked about AI and where it was going, and he pointed to its ability to translate languages. When he said that AI would soon re-place human interpreters, he paused for a moment, realized what he had said, and turned to the interpreter. "Sorry," he said.

The interpreter didn't skip a beat. "Don't worry," she calmly re-plied. "Someone from IBM told me the same thing twenty years ago, and I'm still here."

This conversation illustrates an important point. While it's one thing to predict accurately what jobs AI may replace, it's quite an-other to estimate when computer-based replacements will arrive. Re-peatedly in my quarter century at Microsoft, I've been impressed by the ability of engineering leaders to anticipate much of where com-puting is going. But their predictions around time frames are much more checkered. If anything, people tend to be too optimistic, pre-dicting that change will arrive faster than is the case, but as Bill Gates has famously remarked, "We always overestimate the change that will occur in the next two years and underestimate the change that will occur in the next ten."[15]

This is not a recent phenomenon. Hype around the automobile first took off in 1888, the year Bertha Benz—the wife of inventor Karl Benz, of Mercedes-Benz fame—took her husband's invention and demonstrated to the press what the car could become by driving sixty miles to her mother's house.[16] But when you look at a photo-graph of Broadway in New York taken seventeen years later, in 1905, you'll see a street full of horses and trolleys without a single automo-bile. It takes time for new technology to ripen to the point that it's widely adopted. Yet a photograph taken at the same intersection fif-teen years later, in 1920, shows the street jammed with cars and trol-leys, without a single horse.

The diffusion of new technology seldom occurs at a steady pace. At first, hype outpaces progress, and tech developers need a healthy dose of patience and persistence. Then technology reaches an inflection point, often involving the confluence of several different developments and someone's ability to bring them together in a way that makes the overall product experience more compelling than before. Steve Jobs's success with the launch of the iPhone in 2007 illustrates this well. Mobile phones and handheld personal digital assistants, or PDAs, had each been progressing for a decade. But it was a technical advance in touch-based screens and Jobs's vision for integrating everything with a clean design that led to a rapid explosion of smartphones around the world.

AI likely will be both similar and different. There is good reason to believe that we're reaching a liftoff point for many AI scenarios, like the use of a computer to take an order at a drive-through. But more complex tasks in which errors can result in injury or death—like a self-driving car—may well require considerably more time. As a result, we're likely to see not a single transition across the entire economy or even for a single technology, but rather successive waves and ripples in different sectors. This may characterize technology and societal change over the next two or three decades.

This makes it even more important to think about the cumulative effect of these changes on jobs and the economy. Should we be optimistic or pessimistic about this future? If history offers any insights, which it does, we should be both.

Consider a 2017 McKinsey Global Institute study about the transition to the automobile. It estimated that "the introduction of the automobile created 6.9 million net new jobs in the United States between 1910 and 1950."[17] According to the study, the economy's transition from the horse to the car during these four decades created ten times as many jobs as it destroyed. These jobs represented new occupations that serviced cars and that used motorized vehicles for transportation and delivery.[18] That sounds like powerful cause for optimism.

But here's an opposing reference point. A 1933 report by the US Bureau of the Census, written during the midst of the Great Depression, suggested that the move from horses to cars was "one of the main contributing factors of the present economic situation" affecting the entire country.[19]

How do we explain such diametric conclusions? In one sense, they're both right. In the long run, things have a way of working out. After forty years, the economy had transitioned effectively and the automobile, as well as postwar economic growth, were in full force. But just two decades into this transition, the economy was in dire straits as a result.

Considered from the vantage point of the twenty-first century, it seems difficult to imagine that the transition from horses to automobiles could have had such a negative impact. But it provides insights for our own day from an important and even dramatic backstory that was captured by the Census Bureau, an institution that has always been fueled by data.

In 1933, an agricultural statistician at the bureau named Zellmer Pettet went to work with this data. He had started his career growing fruit in Georgia, where he later joined the Census Bureau as a field agent. While he had earned a college degree in philosophy, he went on to master the intersection of agriculture and what today we would call big data. He authored 115 studies[20] and ultimately retired as the head of the Census of Agriculture.[21] He benefited from the fact that the census takers of the United States didn't just count the number of people living across the country. They also counted the number of horses.

Pettet produced a report about the fire horse's cousin titled "The Farm Horse." While dense with numbers, it offers a compelling story that does for the Great Depression what the 2003 Broadway musical *Wicked* did for the 1939 classic movie *The Wizard of Oz*—it explains many of the events that led to the Depression.

The story begins with the extraordinary degree to which the American economy revolved around horses before the introduction of the automobile. As one historian has commented, "Every family in the United States in 1870 was directly or indirectly dependent on the horse."[22] Across the country, there was one horse for every five people.[23] Because a typical horse consumes ten times as many daily calories as a person,[24] many farmers depended on growing food for horses even more than for people.

Pettet went to work with the Census Bureau's voluminous data and documented what happened after the adoption of the combustion engine. The combination of cars, trucks, and farm machinery between 1920 and 1930 led to a sharp reduction in the nation's horse population—from 19.8 million documented in the 1920 census to 13.5 million a decade later.[25] It was a decline of almost a third. As the horse population fell, so did demand for the food they ate, principally hay, oats, and corn.

The obvious answer was for farmers to switch to growing crops desired by people instead of horses. That is precisely what happened. As Pettet reported, farmers took eighteen million acres of land that had been devoted to horse feed and used it for the production of cotton, wheat, and tobacco.[26] They flooded the market with these crops, depressing their prices. As prices fell, so did the income of farmers. The total revenue earned by farmers from these three crops fell from $4.9 billion in 1919 to $2.6 billion in 1929 and then to only $857 million in 1932.[27] Additional factors contributed to the decline in farm revenue in the early 1930s, but the impact of the declining horse population, while indirect, was both stark and unmistakable.

Soon the nation's rural families found it difficult to pay the mortgages on their farms, leading rural banks to start foreclosing on them. But the banks couldn't keep up with the foreclosures, and soon they had their own problems repaying what they owed to larger banks in the nation's financial centers. In addition, as Pettet found, many jobs in cities were dependent on agriculture-based industries such as packing, manufacturing, and farm machinery.[28] The situation infected the entire country. By 1933, it wasn't just the horse that had lost its job, but almost thirteen million people, or a quarter of the nation's labor force.[29]

As we think about the impact of AI on jobs, what lessons do we draw from almost a century ago? Inevitably, we must be prepared for a roller coaster. There's every reason to expect the transition to an AI era to be as large an upheaval as the transition to the automobile. The demise of the workhorse illustrates the importance of indirect economic effects that are difficult to predict. And inevitably, as in the twentieth century, it will be a transition that calls for innovation not only in technology, but by governments and in the public sector as well. Consider two of the innovations that resulted from the Great Depression: government policies to pay farmers not to overproduce certain crops, and deposit insurance and regulations to ensure the health of banks.

While we can't predict all the areas where new public innovations will be needed, we should assume that such needs will arise. Put in this context, our biggest concern perhaps should not be that technology innovation is so fast, but that government action is so slow. Can democratic governments respond to new needs and crises in an era of political gridlock and polarization? Regardless of where one sits on the political spectrum, this is one of the overarching questions of our time.

There is a second important ramification that can be gleaned from this story. This is the impact of cultural values and broader societal choices on the evolution of technology. It may seem inevitable to us today that cars would replace horses, and there is a lot of truth in this statement from a long-term perspective. But as one author has noted, many of the specific developments were less than inevitable. As she points out, "The replacement of animal power took a particular form that was the result of cultural choices made about energy consumption at the turn of the century."[30] The Progressive movement in the United States championed efficiency, sanitation, and safety improvements in cities, spurring not just the rapid adoption of automobiles that appeared to symbolize these benefits, but the rejection of horse-drawn travel, which created problems in all three areas that urban dwellers understood all too well.

In a similar vein, it would be a mistake to assume that technology trends such as automation and the use of artificial intelligence will be driven by technology and economics alone. Individuals, companies, and even countries will make choices based on cultural values that will manifest themselves in everything from individual consumer preferences to broader political trends that lead to new laws and regulations. These may well diverge in different parts of the world.

There's a final lesson to be drawn from this transition, and it's probably the most encouraging. Just as it's impossible to predict the indirect negative impact of profound technological change, there will be many positive surprises when it comes to the indirect forces that will help create new jobs that don't exist today.

Consider the direct and indirect impacts of the automobile in a place like New York. By 1917, even five years before the fire horse's last day in Brooklyn, New York City was the epicenter of the country's automobile sales. Shops that sold carriages and harnesses on Broadway were replaced by supply stores selling tires and batteries. Where the American Horse Exchange once stood, tall office towers owned by Benz, Ford, and General Motors sprung up. Repair shops, parking garages, gasoline filling stations, and taxi companies were desperate for new, skilled talent to fill their ranks and support America's growing obsession.

None of these direct impacts seem all that surprising. What's more remarkable, even in hindsight, is the rise of new industries that sprang up that at first blush seemed further afield.

A good example was the industry that grew quickly to enable consumer credit. By 1924, 75 percent of cars were paid for over time. Auto installment paper quickly represented over half the country's retail installment credit. Then as now, cars typically were a family's second most expensive possession after purchasing a home. People needed to borrow money to pay for them. As one economic historian has noted, "Installment credit and the automobile were both cause and consequence of each other's success."[31]

This all leads to an interesting question. When New Yorkers saw the first automobile roll down the street in the nation's financial capital, how many predicted that the invention would lead to the creation of new jobs in the financial sector? The route from the combustion engine to consumer credit was indirect and unfolded over time, bolstered in no small part by other intervening inventions and business processes such as Henry Ford's assembly line, which made possible the mass production and thus the cheaper and broader availability of automobiles.

Similarly, the automobile transformed the world of advertising. Seen by passengers traveling in a car at a speed of 30 miles per hour or more, "a sign had to be grasped instantly or it wouldn't be grasped at all," giving rise to the creation of corporate logos that could be recognized immediately wherever they appeared.[32] Yet it's doubtful that the early purchasers of automobiles imagined they would contribute to new jobs on Madison Avenue.

A perspective emerges that is both encouraging and sobering. Technology will make us more productive, relieve us of mundane chores that seem boring, and create exciting new companies and jobs that a future generation will take for granted. It will be an era that will reward those with the determination (and financial capability) to develop new skills and take risks to create new companies. But like the economic impact of the automobile and the cultural loss of the fire horse, we'll undoubtedly suffer setbacks and lose important things in the process. Those who want to slow the pace of technology or avoid outright its negative impacts are likely to be disappointed. The key instead will be to strike a balance between new opportunities and challenges by putting a premium on adaptability, both individually and societally.

In many ways, this is nothing new. People have been adapting to new technology and its impact on jobs since the dawn of the first industrial revolution. It's useful to step back and consider what adaptability has consistently required of people over successive generations. As we thought about what this meant for our own products and future at Microsoft, we concluded that success has always required that people master four skills: learning about new topics and fields; analyzing and solving new problems; communicating ideas and sharing information with others; and collaborating effectively as part of a team.

One goal is to harness AI and create new technology that will help people work better in each of these areas. If we can accomplish that, then we'll provide people with a growing capability not only to confront but to benefit from the next wave of change. Put in this perspective, there is perhaps some room not just for optimism, but even a little faith that human ingenuity will find new ways for people to profit from the technology of tomorrow.

Chapter 14

THE UNITED STATES AND CHINA: A Bipolar Tech World

On a crisp September evening in 2015, a gathering of the who's who from business and government turned out at the Westin Seattle for a lavish banquet. As the dinner course was cleared in the grand ballroom, the 750 people who had come from across the country that evening paused as the guest of honor, dressed in a fine black suit and scarlet tie, took the podium.[1] The room listened intently as the speaker reminisced about his youth, recalled American history, and referenced Western pop culture. He shared tales from his humble background and lifelong love of the works of Ernest Hemingway, Mark Twain, and Henry David Thoreau. He told the crowd how as a student he'd read Alexander Hamilton's *The Federalist Papers*, a collection of articles that was experiencing a bit of a renaissance thanks to the breakout musical *Hamilton*, which had debuted just a month earlier on Broadway.

The speaker wrapped his breezy opening with a reference to his ambition to deliver a better life to his constituents—"the dream," he called it. But the speaker was not your typical American politician.

He wasn't American at all. He was Xi Jinping, the president of China. And the dream he was referring to was "the Chinese Dream."[2]

Standing near former secretary of state Henry Kissinger and current secretary of commerce Penny Pritzker, the Chinese president followed his homespun tale by hitting on the high points we were waiting for in his impressive after-dinner speech, including committing to stopping Chinese cybertheft from American companies and keeping "the door open" to the Chinese market.

Earlier that day, President Xi's plane had touched down on Paine Field, a private airstrip adjacent to the world's biggest factory— Boeing's manufacturing facilities located twenty-two miles north of Seattle in Everett, Washington. It was Xi Jinping's first visit to the United States since he'd become the leader of the most populous country and second-largest economy in the world. His stop in "America's gateway to Asia"[3] was the first in a whirlwind US visit that also included trips to New York and Washington, DC. The historic visit had been months in the making.

The next day, I stood on a red carpet with other Microsoft executives as we waited inside the entrance of Microsoft's Executive Briefing Center. We straightened our ties, double-checked our places in the receiving line, and peered out the glass doors, watching for the Chinese presidential delegation. Every detail of the campus visit had been carefully negotiated and choreographed.

The Chinese government had sent four advance teams during the preceding two months in preparation for Xi Jinping's trip. With each subsequent visit, the group of Chinese planners seemed to double in size. While I had participated in the initial meeting, I hadn't attended the three that followed. A week before the visit, I happened to walk down the hall from my office as the final planning meeting had finished. As I stopped to shake each visitor's hand, I soon realized that I was about to greet more than forty people.

While the logistics mattered, they were trivial compared to the issues that needed to be addressed. Everyone knew that technology would be near the top of the agenda. US companies, Microsoft included, were focused in part on gaining broader access to the Chinese market. In late spring 2015, we had traveled to Beijing to meet with senior Chinese officials to make our case for what we believed would be more open and fair access that would benefit US providers and Chinese customers alike. Slowly we began to see a door open. For the first time in a long time, there was hope.

But just a month later, in early July, the news broke that Chinese hackers had copied from the US Office of Personnel, or OPM, the social security numbers and other personal information of more than twenty-one million Americans.[4] The hackers had penetrated the database containing the details on all Americans with national security clearances. The incident shined a light on both Chinese cybertheft capabilities and the OPM's horrendous lack of security protection.

The next week the White House pulled together a small group to talk with senior administration officials about the incident amid the planning for the upcoming September visit. It was apparent that the cyberhack had understandably touched a nerve in Washington. Officials were not only angry about the data theft, but embarrassed that the hack had been so easy to pull off. That mix of emotions seldom makes for good decision-making.

By late August, the White House team, to its credit, was close to negotiating a new cybersecurity agreement between the two governments, but the situation remained dicey. As planning for the visit progressed, it was apparent that it would make better sense for President Xi to start his trip not in Washington, DC, but somewhere else in the country, creating the opportunity for some positive momentum before he arrived at the White House. The other Washington was a logical choice.

Nine years earlier, President Hu Jintao had made Seattle his first stop on his first official visit to the United States. Bill and Melinda Gates had hosted an elaborate dinner at their house on Lake Washington, and both governments seemed pleased with the outcome. We'd played hosts before and offered to do so again, including a visit to Microsoft. We thought it could provide an incentive to encourage a cybersecurity agreement and create a diplomatic cushion in the event it failed.

As we waited for the massive motorcade to arrive at Microsoft that afternoon, we stood in a carefully arranged order. Satya would greet President Xi first, followed by Bill Gates and John Thompson, the chairman of our board of directors. Then the president would meet me and Qi Lu, the executive vice president who headed our search business and had grown up in China. Satya successfully guided the president on a tour and gave a welcome address, while Harry Shum demonstrated our HoloLens technology.

Then we stepped into a large room for what reporters would call "the most memorable moment" of the state visit—not just at Microsoft or in Seattle, but for the entire six days across the country.[5] The leaders of twenty-eight technology companies from both the United States and China had gathered for a photo opp. Flanking President Xi was a group that included Tim Cook, Jeff Bezos, Ginni Rometty, Mark Zuckerberg, and the CEOs of basically every household technology name in America. It was a photo that built on President Xi's cybersecurity announcement during dinner the night before, one that made every other image of the trip pale in comparison. There was only one president from a country other than the United States who could command this audience. Clearly, President Xi—and the Chinese nation—had taken a central position not just in the global economy, but on the world's technology stage.

China's emergence as a technology superpower, in some respects, signals that we now live in an increasingly bipolar technology world. China and the United States are the world's two largest consumers of information technology. They have also become the two largest suppliers of this technology to the rest of the world. On many days, a scan of stock market listings will show that seven of the world's ten most valuable companies are technology enterprises. Five of these seven are American, while the other two are Chinese. A decade from now, the mix of companies topping this list is likely to have more Chinese firms.

But the technology relationship between the United States and China is unlike anything anywhere else, now or before. While the world previously has seen international IT competition—the United States and Japan vied for leadership in the mainframe era of the 1970s—the dynamic this time is different. In part this is because China has used its size to control access to its market and benefit local providers in a way that no other government could muster. The result is that companies like Google and Facebook, household names everywhere else, are basically absent in China.

While other American companies are present in China, only Apple with its iPhone has enjoyed success in the country on a level that is comparable to its leadership in the rest of the world. In recent years, Apple has earned three times as much revenue as Intel, which is the number two US tech company in China.[6]

When it comes to profits, the situation is likely starker. Apple may well generate more profits within China than the rest of the American tech sector put together. It's a notable accomplishment but also a challenge for the company, given China's large contribution to Apple's global profitability. As we've found at Microsoft over time and more globally with products like Windows and Office, anytime you're dependent on a particular source for a large share of revenue or profitability, it makes it difficult to contemplate changes in that area. It also explains why Apple's leaders are such frequent visitors to Beijing.

More important, Apple's singular success shines a spotlight on everyone else's shortcomings. Why is it so hard for American tech companies to succeed in China compared with the rest of the world? This has been a leading question across the tech sector for more than a decade. And increasingly in Washington, DC, politicians in both parties are asking whether they want US tech companies to succeed there, given the potential technology transfer involved.

The technology relationship between the United States and China has emerged as the most complex in the world—and probably in history.

As competition grows, it's critical for each country to understand the other. Too often the history of international relations has been marked by views of other countries that are based more on caricatures than true understanding. There are multiple reasons American companies have encountered more difficulty in China than in other places. It's important to put them together in context.

One thing that has become increasingly apparent is that Chinese consumers sometimes have different information technology needs and interests than consumers in the United States, Europe, and elsewhere. American tech companies, Microsoft included, typically have brought to the Chinese market products designed initially for users in the United States. At times, these products have met the needs and satisfied the tastes of Chinese users. Hardware like the iPhone and Microsoft Surface and productivity software like Microsoft Office are good examples. But at other times, Chinese users are drawn to entirely new and different approaches.

Bill Gates had the foresight to recognize more than two decades ago that China would emerge not only as a large market of customers but also as an important country for technology talent. In November

1998, the doors opened at Microsoft Research Asia, or MSRA, now located within two towers in Beijing close to two leading academic institutions—Tsinghua and Peking universities. In its first two decades, MSRA's pioneering research has focused on not just computer science fundamentals, but a wide variety of fields, including natural language and natural user interfaces, data-intensive computing, and search technologies.[7] Its researchers have published more than fifteen hundred academic papers in fields that have contributed to computer science advances around the world. MSRA is emblematic of China's rapidly growing technology talent base.

At times, MSRA moves beyond basic research to develop and pilot new products designed specifically for the Chinese market. From an American perspective, these are sometimes surprising. One example is a product called XiaoIce, a female AI-based social chatbot designed to have conversations with teenagers and people in their early twenties.[8] The chatbot seems to have filled a social need in China, with users typically spending fifteen to twenty minutes talking with XiaoIce about their day, problems, hopes, and dreams. Perhaps she fills a need in a society where children don't have siblings? This social chatbot has grown to serve more than six hundred million users, and her capabilities are growing, including AI-based applications to compose poems and songs. XiaoIce has become a celebrity of sorts; she has made a guest appearance as a television weather forecaster and regularly hosts TV and radio programs.[9]

The world's different tastes in technology were revealed when we brought XiaoIce to the United States in the spring of 2016. We launched her to the US market under the name Tay. The new name turned out to be just the start of our problems with XiaoIce's American debut.

I was on vacation when I made the mistake of looking at my phone during dinner. An email had just arrived from a Beverly Hills lawyer who introduced himself by telling me, "We represent Taylor Swift, on whose behalf this is directed to you." That opening alone set the email apart from the rest of my inbox. He went on to state that "the name 'Tay,' as I'm sure you must know, is closely associated with our client." No, I actually didn't know, but the email nonetheless grabbed my attention.

The lawyer went on to argue that the use of the name Tay created a false and misleading association between the popular singer and our chatbot, and that it violated federal and state laws. Our trademark lawyers took a different view, but we hadn't sought to pick a fight with or even offend Taylor Swift. There were plenty of other names we could choose, and we quickly talked about finding a substitute.

Almost immediately, however, we had bigger issues to worry about. Tay, like XiaoIce, could be trained to interact with people based on feedback in conversations. A small group of American pranksters had organized an effective campaign using Tweets to train Tay to utter racist comments. In little more than a day we had to withdraw Tay from the market to address the problem, providing a lesson not just about cross-cultural norms but about the need for stronger AI safeguards.[10]

Tay was but one example of differing cultural practices across the Pacific. Services developed in the United States have fallen flat because Chinese users prefer different products with different approaches developed in and for their own country. Much more notable has been the triumph of Chinese services like Alibaba over Amazon in e-commerce, Tencent's WeChat over American services in messaging, and Baidu over Google in search. In important and well-documented respects, these services innovated to meet Chinese tastes in ways that their American counterparts did not.

This underscores a tech trait encountered increasingly around the world, especially in China. There are smart people everywhere, and Chinese companies are innovating, working hard, and succeeding in substantial measure based on a commitment to innovation and the strong work ethic long valued by those who champion free enterprise, including in the United States. You see this not just in the Chinese companies creating technology tools, but in the entities across Chinese society that are now deploying AI-based advances at a remarkable rate. This rapid growth in market deployment is adding fuel to the impressive engines of Chinese technology. Put all together, it's helping China's tech sector create more formidable local competition than American tech companies have encountered anywhere else.

But other factors hampering our success are more challenging. They start with barriers to market access in China—and increasingly in the United States.

In the race to construct barriers for technology market access, China indisputably was the early global leader. It's not that other countries weren't tempted. It's just that the price of participation in the world trading system, especially through the World Trade Organization, precluded this strategy. Along with a sustained and determined focus by the US Trade Representative, a combination of multilateral and bilateral negotiations opened markets for the American tech sector around the world.

China alone has had the market size and determined strategy to resist this approach. Products that could be freely imported elsewhere required one or more complicated government licenses before being made available in China. Even when licensed, American tech companies often have found that the Chinese public sector and other large customers will purchase and use the technology only if it's offered through a joint venture with a Chinese partner.

In the best of times, joint ventures in the tech sector have been notoriously difficult to make work. Information technology changes rapidly and often involves substantial engineering complexity. Business models tend to evolve as well, and all this creates the need for ongoing changes in marketing, sales, and support. In an industry where large acquisitions often fail, joint ventures fare even worse. And that's before one adds the complexity of working across countries, cultures, and languages.

Even an informal obligation to enter the market through a joint venture is like requiring a cross-country runner to race with a backpack full of heavy weights. It's a rare day when one wins such a race, and the odds are even more daunting when competing with impressive local companies unhindered in a similar way. In short, an obligation to serve China through a joint venture operates as a real and typically effective barrier to market access.

But the technology issues between China and the United States go well beyond market access. Given the role that information technology plays in broad communication, unfettered expression, and social movements, the Chinese government has long regulated its use in a way that is different than in the West. For any American tech company, entry into the Chinese market requires what often feels like a dizzying array of constantly evolving regulations from myriad government agencies at both the national and provincial levels. There are many difficult days that turn on thorny issues with clear tensions between a Chinese focus on public order and a Western commitment to human rights.

In some respects, these differences are rooted in even deeper contrasts in philosophy and views of the world. It's important to understand all these issues and how they fit together.

As University of Michigan professor Richard Nisbett noted in his 2003 book, *The Geography of Thought*,[11] these issues reflect differing and deep philosophical traditions that go back more than two thousand years. American thought is often based in part on philosophies developed in ancient Greece, while Chinese thinking is founded on the teachings of Confucius and his followers. Over two millennia these have emerged as two of the world's most prevalent and influential—but also different—ways of thinking.

While I've spent decades attending meetings around the world, Beijing stands alone as a capital where government discussions sometimes hark back explicitly to historical experiences that date back over two millennia. Or to be specific, to 221 BC, the year the Qin dynasty unified China.

As Henry Kissinger has noted, China owes its millennial survival to "the community of values fostered among its population and its government of scholar-officials."[12] Kissinger has probably spent more time focused on China than any other American official of the past century. As he observes, the values that continue to guide official thinking in China today derive from teachings by Confucius, who died more than two centuries before the Qin dynasty was born. His teachings included a commitment to compassionate rule, devotion to learning, and a pursuit of harmony based on a hierarchical code of social conduct that included a fundamental duty to "Know thy place."[13]

As Nisbett points out, the Greek philosophy that remains the foundation of Western political thought shared a strong sense of curiosity with Confucius's devotion to learning, but was grounded in a different sense of personal agency—a sense that people "were in charge of their own lives and free to act as they chose."[14] As developed by Aristotle and Socrates, the very definition of happiness for the ancient Greeks "consisted of being able to exercise their powers in pursuit of excellence in a life free from constraints."[15]

As a company founded and headquartered in the United States, we didn't doubt our own historical roots or the importance of protecting human rights around the world. We decided a decade ago not to host our consumer email on servers in China because of the human rights risks that this would have created, even though the Chinese government made clear that this meant our service would no longer be available to the country's consumers. And I'll always remember the late-night phone calls when I've insisted that employees on the front line in China hold firm on what we concluded were unlawful censorship demands on our search service, holding my breath as I sat in the comfort of my home knowing that they were sitting uncomfortably having to relay my answer to local officials. More recently, we restricted access to our facial-recognition services, given the potential for mass surveillance.

Episodes like these called the question in concrete terms regarding our commitment to human rights. While we've long been dedicated to supporting our customers and growing in China, we also concluded that it was imperative to approach this in a principled manner. In ways that have become even clearer over time, it has been essential to stay grounded in an approach that puts the priority on fundamental values, including universal human rights, over revenue growth and the bottom line.[16]

From our perspective, these fundamental differences also make it even more important for people in the world's two largest economies to learn more about each other's cultures and historical traditions. While it's easy for one country or the other to turn away, this will not make any of these differences go away.

We had the opportunity to learn more about these differences in Beijing first-hand in the summer of 2018. We had arrived early for a week in Asia to spend a sweltering Sunday digging into the connection and contrasts between the most modern of technologies—AI—and the philosophical and religious traditions that emerged over literally thousands of years.

The Microsoft team and I started our morning at the Longquan

Temple, a compound of multistory stone-and-wood buildings capped with sweeping Buddhist rooflines. Nestled in what locals call the lung of the city—the lush area of Phoenix Mountain, a nature park in the western outskirts of Beijing—the monastery was founded during the Liao dynasty. It's a peaceful place, straddling a hillside stream and home to thousands of humming cicadas. We walked along the meandering paths and through the gardens with interest, but nothing delighted us more than when our host showed us the AI projects he'd been working on.

As Master Xianxin explained to us, the monastery had devoted itself to merging Buddhist teachings and traditions with the modern world. He was a graduate of the Beijing University of Technology. Yes, a Buddhist monk with a computer science degree. He showed off thousands of volumes of ancient Buddhist literature the temple was digitizing with the help of AI. The master went on to share how the monks were using machine-based translation techniques to share their work in sixteen languages with people around the world. Modern technology was advancing some of the world's most ancient teachings.

Later that afternoon, we traveled into the center of Beijing to meet with a professor named He Huaihong, one of the country's leading philosophers and ethicists. Professor He was based at Peking University and had published a book about changing social ethics in China.[17] Even a cursory reading of the book dispels the notion that, at least in some areas, there is an absence of vibrant debate in contemporary China.

We talked about the ethical and philosophical issues presented by AI and how they might be viewed differently in various parts of the world. It was striking when one of Professor He's first comments echoed some of the opening words in Nisbett's book written fifteen years before. "In the West, there's more of a belief in progress as a straight line, with technology moving forward and optimism about constant improvement," he said.

As Nisbett had noted, people in the West tended to focus on a specific goal and believed that if you could pour yourself into advancing it, you could change the world around you. It was part of the entrepreneurialism that made Silicon Valley not just a location but an attitude that fueled innovation.

"In China," Professor He said, "we see everything moving in cycles. Like the signs of the zodiac, we believe that life is a circle and that everything will come back to its original point at some time in the future." It led people in China to look backward as well as forward and to focus more on the whole picture rather than an individual piece.

As Nisbett had explained, the Pacific is indeed a wide ocean when it comes to how people on each side tend to look at the same image. Take a photograph of a tiger in the jungle. Americans are more likely to focus on the tiger and what it can do. The Chinese are more likely to focus instead on the jungle and the way it influences every aspect of the tiger's life. Neither approach is wrong, and arguably a combination of the two could be most valuable. But the differences are clear.

These different traditions are also consistent with how each society thinks about new technology and the regulation of it. Americans' instinct is to keep government at a distance, so a young "tech tiger" can grow up, change, and become stronger, with optimism about what it will achieve. The Chinese are far quicker to address the "societal jungle" that the tech tiger inhabits, including by imposing a web of government regulations that govern the tiger's activities.

This is an added dimension that helps explain the complicated relationship between technology companies and the government in China. There is much more than a language barrier that needs to be overcome. Tech companies have worked together and with the global human rights community to encourage adherence to global principles relating to privacy and free expression. But on some days these principles receive a less global endorsement than when they were endorsed by the world's governments, including the Chinese, shortly after the end of World War II. There are moments that involve complicated discussions that, at their foundation, feel like a negotiation not only around political approaches, but about the alternative worldviews of Aristotle and Confucius.

If these differing philosophical lenses don't make for enough complexity on their own, the cybersecurity issues of the past decade have added more challenges. The US government understandably reacted strongly not only to incidents such as the hacking of the OPM but also to reports that Chinese hardware manufacturer Huawei had built routers that enabled the Chinese government to monitor communications by customers who used them.[18] The shoe then was on the other foot when the Snowden disclosures included a photo of US personnel tampering with Cisco routers to achieve the same thing.[19] Both companies have been working since—with less than complete success—to recover their reputations in the other's market.

Increasingly, in Washington, DC, both political parties have viewed the rise of Chinese influence with concern. While President Trump has pushed China hard to increase its purchase of almost every American product, one category has had more qualified support: information technology. Believing that this technology will be increasingly fundamental to both economic strength and military power, American policy makers have expressed increasing concern about the prospects of ongoing technology transfers to China.

While these clearly are important and broadening concerns, there's a risk on both sides of the Pacific in applying simple answers to complex questions. In both countries, there are important nuances that need to be considered.

To start with, there are some information technologies that are sensitive from a national security or military perspective, but there are many others that are not. And the idea that some technology could be useful for both military and peaceful purposes is far from new. Such "dual use" products have been around for decades, and a well-established export-control regime exists to govern them. Nonetheless, there's a growing risk that American policy makers will fail to take into account some of the vital differences between information technology and other technologies that are important to national security as they consider the ongoing rise of China.

In addition, while there is some information technology that is secret, there's a lot that is not. Unlike many military technologies, advances in computer and data science often take place at the basic research level and are published initially in the form of academic papers. They're available to the world. In addition, software almost uniquely often consists of source code that is published in open-source form, meaning that anyone anywhere can not only read it, but incorporate it into their own products. While a concern around the protection of trade secrets is important and well placed in some computer science fields, in some software areas trade secrets have little practical application.

There are also some technology scenarios that clearly raise human rights concerns while others do not. Facial-recognition services and citizen and consumer data stored in the cloud are two that do. On the other hand, since the 1980s we've distributed Microsoft Word so users can run it on their own computers without anyone else knowing what they were writing. While Word Online now runs in the cloud, people can choose which version they want to use and how they want to use it. Even in the context of human rights, the same software can have dramatically differing impacts in different scenarios.

Finally, China itself is a vital part of the supply chain for

American technology products. At one level, this is generally understood when it comes to the manufacturing of components for computer hardware. But China's role goes well beyond this. The country's burgeoning number of engineers is integrated into a global process for research and development. Most technology companies incorporate research advances developed by Chinese engineers together with advances by engineers in the United States, the United Kingdom, India, and many other places around the world. While policy makers might contemplate creating a new iron curtain down the middle of the Pacific Ocean to separate technology development on separate continents, the global nature of technology development makes this difficult to implement. And even if such a barrier is erected, it's far from clear whether a country that pursues such an approach will benefit or simply slow its own technology development.

All this means that the United States and China each confront a growing conundrum about how to think about technology trade. There are three long-term dimensions that need to be considered.

First, on the import side, one would be hard-pressed today to say that either American or Chinese technology companies have unfettered access to each other's markets. To the contrary, there has emerged something of a home court advantage for IT leaders in each country. One result is that US and Chinese companies increasingly succeed more easily at home and compete in the rest of the world.

From an international economic perspective, it's worth remembering that, for the companies involved, this protection of home markets is a blessing that has its share of curses. Even for China, with its 1.4 billion people, more than 80 percent of the world's consumers live and work somewhere else. The only way to succeed globally as a technology leader is to be respected globally. Both American and Chinese technology companies share a need to win over customers outside their borders when they're seeking to grow in Europe, Latin

America, or across the rest of Asia or elsewhere around the world. If the US and Chinese governments each allege that the other country's homegrown technology can't be trusted, there's a risk that the rest of the world will conclude that both are right and turn toward other sources.

At one level, there are clear sensitivities around network components like 5G products that are foundational for national infrastructure in times of peace and war alike. Given not just the potential for but also the history of nation-state tampering and hacking, a focus on this area is understandable. But even in this space, it's vital for national policies to be well grounded in objective facts and logical analysis. Governments should be even more circumspect and careful when considering tactics that make a point through criminal prosecutions or other serious legal actions against specific firms or individuals.

Looking beyond 5G, steps to shut out long lists of technology services in many other areas are likely to be both unnecessary and counterproductive. There exist plenty of other ways to regulate most technology services in a reliable and country-neutral way, assuming regulation is needed at all. If anything, it's in the economic self-interest of the world's two technology leaders to keep most of their technology markets open to others, thereby setting an example for the rest of the world to emulate.

Second, there's a growing focus on the export side of the trade equation, especially in Washington, DC. And this is creating an increasing possibility that American officials will seek to block the export of a growing number of vital technology products, not just to China but to a growing set of other countries.

The risk here is that US officials will fail to appreciate that technology success almost always requires success on a global scale. The economics of information technology turn on spreading R&D and infrastructure costs over the largest number of users possible. This is what drives down prices and creates the network effects needed to turn new applications into market leaders. As LinkedIn cofounder (and Microsoft board member) Reid Hoffman has shown, the ability to "blitzscale" quickly to global leadership is fundamental to technology success.[20] But it's impossible to pursue global leadership if products can't leave the United States.

All this makes a new generation of potential US export controls even more challenging than in the past. It argues for both proceeding with caution and considering new export approaches. In the past, export control officials worked from lists of products that sometimes were banned from the export market entirely. For many emerging technologies from AI to quantum computing, it likely makes more sense to permit certain technologies to be exported but with limitations that restrict their availability for certain uses and users. While this will inject greater complexity into export administration for governments and companies, it may well be the only way to protect national security while promoting economic growth.

Finally, there are broader dimensions that need to be considered, not just for the United States and China, but for the world. The two nations increasingly are dividing the internet almost in half when it comes to the global population's use of technology. Even more broadly, it's almost impossible to imagine this century ending in a better state than it began without a healthy relationship across the Pacific. Put simply, the world needs a stable relationship between the United States and China, including on technology issues.

This requires the continued building of a stronger educational and cultural foundation to connect the United States and China. The technology issues for the two countries require a common comprehension not only of science and engineering but also of language, the social sciences, and even the humanities. Today each country's understanding of the other is often more limited than it should be.

In most respects, this limitation is the strongest in the United States. Consider the fact that President Xi's education included reading American authors from Alexander Hamilton to Ernest Hemingway. How many American politicians have read comparable Chinese authors? With more than twenty-five hundred years of rich history, the problem is not a lack of supply but a shortage of interest. As history has demonstrated repeatedly, if the United States is going to navigate global challenges, it will need leaders who understand the world.

Ultimately the United States and China need a bilateral relationship that serves the interests of each country. Each nation's leaders rightly will focus on their own self-interest with eyes wide open and in a hardheaded way. But whenever the governments of the two largest economies in the world get together, there's a responsibility to think not only about their own individual and collective interests but also about what their relationship means for the rest of the world. Increasingly the rest of the world—with almost 80 percent of the global population—is depending on it.

Chapter 15

>>> **DEMOCRATIZING THE FUTURE: The Need for an Open Data Revolution**

What impact will data and AI have on the distribution of geopolitical power and economic wealth? It's another dynamic that centers in part on the United States and China, but with even broader implications for the rest of the world. It's one of the overarching questions of our time, and in the fall of 2018 a pessimistic view emerged.

As we met with members of Congress in Washington, DC, a few senators mentioned that they had read the advance galleys sent to them for a new book called *AI Superpowers*. Its author, Kai-Fu Lee, is a former executive at Apple, Microsoft, and Google. Born in Taiwan, he is now a leading venture capitalist based in Beijing. His argument is sobering. He asserts that "the AI world order will combine winner-take-all economics with an unprecedented concentration of wealth in the hands of a few companies in China and the United States."[1] As he puts it, "other countries will be left to pick up the scraps."[2]

What's the basis for this view? Mostly it comes down to the power of data. The argument is that the firm that gains the most users will gain the most data, and because data is rocket fuel for AI, its AI product will become stronger as a result. With a stronger AI product, the firm will attract even more users and hence more data. The cycle will continue with returns to scale, so that eventually this firm will crowd out everyone else in the market. According to Kai-Fu, "AI naturally gravitates toward monopolies . . . once a company has jumped out to an early lead, this kind of ongoing repeating cycle can turn that lead into an insurmountable barrier to entry for other firms."[3]

The concept is a common one in information technology markets. It's referred to as "network effects." It has long been true in the development of applications for an operating system, for example. Once an operating system is in a leadership position, everyone wants to develop apps for it. While a new operating system might emerge with superior features, it's difficult to persuade app developers to consider it. We benefited from this phenomenon in the 1990s with Windows and then hit the barrier on the other side twenty years later, competing against the iPhone and Android with our Windows Phone. Any new social media platform that wants to take on Facebook encounters the same problem today. It's part of what defeated Google Plus.

According to Kai-Fu, AI will benefit from a similar network effect on steroids, with AI leading to increased concentration of power in nearly every sector of the economy. The company in any sector that most effectively deploys AI will gain the most data about its customers and create the strongest feedback loop. In one scenario, the outcome could be even worse. Data could be locked up and processed by a few giant tech companies, while every other economic sector relies on these companies for their AI services. Over time, this would likely lead to an enormous transfer of economic wealth from other industrial sectors to these AI leaders. And if, as Kai-Fu projects, these companies are located mostly on the east coast of China and the west coast of the United States, then these two areas will benefit at every other region's expense.

What should we make of these predictions? Like many things, they are based on a kernel of truth. And in this case, perhaps more than one.

AI depends upon cloud-based computing power, the development of algorithms, and mountains of data. All three are essential, but the most important of these is data—data about the physical world, the economy, and how we live our daily lives. As machine learning has evolved rapidly over the past decade, it has become apparent that there's no such thing as too much data for an AI developer.

The implications of data in an AI-driven world go well beyond the impact on the tech sector. Consider what a product like a new automobile will be like in 2030. One study recently estimated that fully half of the cost of a car by that time will consist of electronics and computing components, up from 20 percent in the year 2000.[4] It's apparent that by 2030, cars will always be connected to the internet for autonomous or semiautonomous driving and navigation, as well as for communications, entertainment, maintenance, and safety features. All of this is likely to involve artificial intelligence and large quantities of data based on cloud computing.

This scenario raises an important question: What industries and companies will reap the profit generated from what increasingly is a massive AI-driven computer on wheels? Will it be traditional automobile makers or tech companies?

This question has profound implications. To the extent this economic value is retained by automakers, there is cause for more optimism in the longer-term futures of car companies like General Motors, BMW, Toyota, and others. And, of course, this is likely to provide brighter prospects for the salaries and jobs at these companies and for the people who hold them. Put in this context, it's apparent that this is also an important issue for these companies' shareholders and for the communities and even nations where these companies are based. It's not an exaggeration to say that the economies in places like Michigan, Germany, and Japan have their future riding on this outcome.

If this seems far-fetched, consider Amazon's impact on book publishing—and now many retail sectors—or what Google and Facebook have done to advertising. AI could have the same impact on everything from airlines to pharmaceuticals and shipping. This in effect is the future painted by Kai-Fu Lee. It's why there is at least a plausible basis to conclude the future could bring an ever-increasing transfer of wealth to the small number of companies that hold the largest pools of data and the regions where they are based.

As is so often the case, however, there is no single and inevitable path into the future. While there is a risk the future could unfold in this manner, there is an alternative course we can chart and pursue. We need to empower people with broader access to all the tools needed to put data to work. We also need to develop data-sharing approaches that will create effective opportunities for companies, communities, and countries large and small to reap the benefits from data. In short, we need to democratize AI and the data on which it relies.

So how do we create a bigger opportunity for smaller players in a world where large quantities of data matter?

One person who may have the answer is Matthew Trunnell.

Trunnell is the chief data officer at the Fred Hutchinson Cancer Research Center, a leading cancer research center in Seattle named for a hometown hero who pitched ten seasons for the Detroit Tigers and managed three major league baseball teams. In 1961, Fred Hutchinson took the Cincinnati Reds to the World Series.

Sadly, Fred's successful baseball career and life were cut short when he died of cancer in 1964 at the age of forty-five.[5] His brother, Bill Hutchinson, was a surgeon who treated Fred's cancer. After his younger brother's death, Bill founded the "Fred Hutch," a research center devoted to curing cancer.

Trunnell came to Seattle in 2016 to work at the Hutch. The Institute has twenty-seven hundred employees working in thirteen buildings that sit on the south shore of Lake Union. Seattle's iconic Space Needle is visible in the distance.

The Hutch's mission is ambitious: to eliminate cancer and its related deaths as a cause of human suffering.[6] It brings together scientists, three of whom have won Nobel Prizes, with doctors and other researchers to pursue cutting-edge research and treatments. It partners closely with its neighbor, the University of Washington, which has globally renowned medical and computer science centers. The Hutch has built an impressive track record that has included innovative treatments for leukemia and other blood cancers, bone marrow transplants, and now new immunotherapy treatments.

The Hutch has become like almost every institution and company in virtually every field on earth: Its future depends on data. As Hutch president Gary Gilliland has concluded, data is "going to transform cancer prevention, diagnosis and treatment."[7] He notes that researchers are turning data into a "fantastic new microscope" that shows "how our immune system responds to diseases like cancer."[8] As a result, the future of biomedical science is no longer in biology alone, but in its convergence with computer science and data science.

While Trunnell has never met Kai-Fu Lee, this recognition set him on a path that in effect challenges the author's thesis that the future belongs only to those who control the world's largest supply of data. If that were the case, then it would be difficult for even a world-class team of scientists in a midsize city in a far corner of North America to aspire to be among the first to find a cure for one of the planet's most challenging diseases. The reason is clear. While the Hutch has access to important collections of health record data that help it pursue AI-based cancer research, in no way does it possess the world's largest data sets. Like most organizations and companies, if the Hutch is to continue to lead into the future, it must compete without actually owning all the data that it will need.

The good news is that there is a clear path to success. And it builds on two features that set data apart from most other important resources.

First, unlike traditional natural resources such as oil or gas, humans create data themselves. As Satya put it during one of our Friday senior leadership team meetings at Microsoft, data is probably "the world's most renewable resource." What other valuable resource do we create many times unintentionally? Human beings are creating data at a rapidly growing rate. Unlike resources for which there is a finite supply or even a shortage, the world if anything is awash in an ever-expanding ocean of data.

This doesn't mean that scale doesn't matter or that larger players don't have an advantage. They do. China has more human beings and hence more capacity to create data than any other country. But unlike, say, the Middle East, which has more than half of the world's proven oil reserves,[9] it will be hard for any country to corner the world's market on data. People everywhere create data, and over the course of this century, it seems reasonable to expect nations everywhere to create data in some rough combination of their relative population size and economic activity.

China and the United States may be early AI leaders. But China, as large as it is, accounts for only 18 percent of the world's population.[10] And the United States represents only 4.3 percent of the world's people.[11] When it comes to the size of their economies, the United States and China have more of an advantage. The US represents 23 percent of the world's GDP, while China represents 16 percent.[12] But because the two countries are far more likely to compete than join forces, the real question is whether one nation can dominate the world's data with less than a quarter of the global supply.

While there is no single guaranteed outcome, there is a greater opportunity for smaller players based on data's second feature, which, as it turns out, is even more critical. Data, as economists put it, is "non-rivalrous." When a factory is powered by a barrel of oil, that barrel is not available to any other factory. But data can be used again and again, and dozens of organizations can draw insights and learning from the same data without detracting from its utility. The key is to ensure that data can be shared and used by multiple participants.

Perhaps not surprisingly, the academic research community has been a leader in using data in this way. Given the nature and role of academic research, universities have begun to set up data depositories, where data can be shared for multiple uses. Microsoft Research is pursuing this data-sharing approach too, making available a collection of free data sets to advance research in areas such as natural language processing and computer vision, as well as in the physical and social sciences.

It was this ability to share data that inspired Matthew Trunnell. He recognized that the best way to accelerate the race to cure cancer is to enable multiple research organizations to share their data in new ways.

While this sounds simple in theory, its execution is complicated. To begin, even in a single organization data is often stashed in silos that must be connected, a challenge made even greater when the silos sit in different institutions. The data may not be stored in a form that is readable by machines. Even if it is, different data sets are likely to be formatted, labeled, and structured in different ways that make it harder to share and use in common. If the data came from individuals, legal issues around privacy will need to be worked through. And even if the data doesn't involve personal information, other big questions need to be hammered out, such as the governance process among organizations and the ownership in data as it grows and is improved.

These challenges are not just technical in nature. They're also organizational, legal, social, and even cultural. As Trunnell recognized, they stem in part from the fact that most research institutes have conducted much of their technology work with tools they developed in-house. As he says, "In addition to keeping data siloed at one organization, this approach often results in duplicative data collection, lost patient histories and outcomes, and a lack of knowledge about potentially complementary data elsewhere. Together these problems hinder discovery, slow the pace of health data research, and drive up costs."[13]

The collective impact of all these impediments, Trunnell observed, makes it difficult for research organizations and technology companies to partner with each other. And the result, he found, hinders the aggregation of data sets large enough to support machine learning. In effect, the inability to overcome these barriers offers the best prospect for the AI dominance envisioned by Kai-Fu Lee.

Trunnell and others at the Hutch saw a data problem that needs to be solved, and they set out to solve it. In August 2018, Satya, himself a Hutch board member, invited a group of senior Microsoft employees to a dinner to hear about the Hutch's work. Trunnell spoke about his vision for a data commons that would enable multiple cancer research institutes to share their data in new ways. His vision would bring together several organizations to pool their data in partnership with a tech company.

My enthusiasm grew as I listened to his presentation. In many ways, the challenge was like many others we had learned about and even experienced ourselves. As Trunnell described his plans, it reminded me of the evolution of software development. In the early days of Microsoft's history, developers protected their source code as a trade secret, and most tech companies and other organizations developed their code by themselves. But open source had revolutionized the creation and use of software. Increasingly software developers were publishing their code under a variety of open-source models that allowed others to incorporate, use, and contribute improvements to it. This enabled broad collaboration among developers that helped accelerate software innovation.

When these developments began, Microsoft had not only been slow to embrace the change, we'd actively resisted it, including by asserting our patents against companies shipping products with open-source code. I had been a central participant in the latter aspect. But over time, and especially after Satya became the company's CEO in 2014, we began to recognize this was a mistake. In 2016, we acquired Xamarin, a start-up that supports the open-source community. Its CEO, Nat Friedman, joined Microsoft and brought an important outside perspective to our leadership ranks.

By the start of 2018, Microsoft was using more than 1.4 million open-source components in its products, contributing back to many of these and other open-source projects, and even open sourcing many of its own foundational technologies. As a sign of how far we'd come, Microsoft had become the most prolific contributor to open source on GitHub,[14] a company that was the home of software developers around the world and especially of the open-source community. In May, we decided to spend $7.5 billion to acquire GitHub.

We decided that Nat would lead the business, and as we worked on the deal, we concluded that we should join forces with key open-source groups and do the opposite of what we had done a decade before. We would pledge our patents to defend the open-source developers that had created Linux and other key open-source components. As I talked this through with Satya, Bill Gates, and then others on our board of directors, I said it was time to "cross the Rubicon." We had been on the wrong side of history, and as we all concluded, it was time to change course and go all in on open source.

I recalled these lessons as I listened to Trunnell describe the data commons. The challenges, while complicated, were like many the open-source community had overcome. Within Microsoft, our increasing use of open-source software had led us to think through the technical, organizational, and legal challenges involved in creating it. More recently we had built one of the tech sector's leading efforts to address the privacy and legal challenges in working on shared data use as well.

But even more striking than the pitfalls was the promise of what Trunnell described. What if we could create an open-data revolution that would do for data what open-source code had done for software? And what if this approach can outperform the work of an inward-facing institution relying on the largest proprietary data set?

The discussion reminded me of a meeting I had attended a couple of years earlier, which had surprisingly ended up focusing on the real-world impact of sharing data.

In early December 2016, a month after the presidential election, a meeting was held in Microsoft's Washington, DC, offices to examine the impact of technology on the presidential race. The two political parties and various campaigns had used our products, as well as technology from other companies. Groups of Democrats and Republicans had agreed to meet with us separately to talk about how they'd used technology and what they'd learned.

We first met with advisers on the team from Hillary Clinton's campaign. Throughout the 2016 campaign season, they were considered the country's political data powerhouse. They had set up a large analytics department that built on the success of the Democratic National Committee, or DNC, and Barack Obama's successful campaign for reelection in 2012.

The Clinton campaign had leading tech experts building what was regarded as the world's most advanced campaign tech solutions to utilize and improve what was perhaps the country's single best political data set. As the tech and campaign advisers told us, Robby Mook, the bright and affable Clinton campaign manager, had based most of his decision making on the insights generated by the analytics department. Reportedly as the sun set on election day on the East Coast, the entire campaign organization believed they had won the race, thanks in no small measure to their data analysis capabilities. At about dinnertime, the analytics team stepped away from its computers to receive a standing ovation from a grateful campaign staff.

A month later, that initial applause had been replaced by a growing silence from the analytics side of the defeated Clinton campaign. The campaign team had been criticized publicly for missing a rising Republican swing in Michigan until a week before the election and in Wisconsin until the night the votes were counted. But there still prevailed a high degree of confidence in the campaign's data prowess. As our debrief reached its conclusion, I asked the Democratic team assembled a simple question: "Do you believe you lost because of your data operation or in spite of it?"

Their reaction was both swift and full of self-confidence. "Without question we had the better data operation. We lost despite that."

We took a break as the Democratic team left, and a team of leading Republicans sat down with us to compare notes.

As they described the course of the campaign, the surprising twists and turns that had led to the nomination of Donald Trump had a decisive impact on his campaign's data strategy. Shortly after Barack Obama's reelection in 2012, Reince Priebus was reelected to a second term heading the Republican National Committee, or RNC. He and his new chief of staff, Mike Shields, undertook a top-to-bottom review of the RNC's operations in the wake of the 2012 defeat, including its technology strategy. And as often happens in the fast-paced world of technology, there emerged an opportunity to leapfrog the competition.

Priebus and Shields utilized data models from three Republican technology consulting firms and embedded them in-house at the RNC. While they lacked easy access to the Democratic-leaning talent pool in Silicon Valley, they brought in a new CTO from the University of Michigan and a young technologist from the Virginia Department of Transportation to build new algorithms for the world of politics. The two RNC leaders believed—and proved—that there was top data science talent everywhere.

Most important to the Republican tech strategists that morning was what Preibus and his team did next. They succeeded in establishing a data-sharing model that convinced not only Republican candidates across the country but also a variety of super PACs and other conservative organizations to contribute their information to a large, federated file of foundational data. Shields believed it was important to assemble as much data as possible from as many sources as possible in part because the RNC had no idea who the ultimate presidential nominee would be. Until then, they couldn't know what types of issues or voters the candidate would find most important. So the RNC team worked to connect with as many organizations and to federate as much diverse data as possible. It created a much richer total data set than anything the DNC or the Clinton campaign possessed.

When Donald Trump secured the Republican nomination in the spring of 2016, his operation lacked the deep technology infrastructure of the Clinton campaign. To make up for this deficit, Trump's son-in-law, Jared Kushner, worked with the campaign's digital director, Brad Parscale, on a digital strategy that would build on what the RNC already had rather than create their own. Based on the RNC's data sets, they had identified a group of fourteen million Republicans who said they did not like Donald Trump. To turn this group of skeptics into supporters, the Trump team created Project Alamo in Parscale's hometown of San Antonio to consolidate fund-raising, messaging, and targeting, especially on Facebook. They communicated to these voters repeatedly with messages on topics that the data said were likely to be important to them, like the opioid epidemic and the Affordable Care Act.

The Republican team described what their data operation revealed as the election approached. Ten days before the election, they estimated that they were down two points to Clinton in key battleground states. But they had identified 7 percent of the population that was still undecided about whether it would vote. And the campaign had email addresses for seven hundred thousand people who, the team believed, were likely to vote for Trump in these states if they went to the polls. They put all their energy into persuading this group to turn out.

We asked the Republican team what technology lessons they had learned from their experience. Those lessons were several: Don't go as deep as the Clinton team had gone in building a data operation from the ground up. Instead use one of the major commercial tech platforms and focus on building on top of it. Build with a broader federated ecosystem that brings together as many partners as possible to contribute and share data, as the RNC had done. Use this approach to focus resources on differentiated capabilities that can run on top of a commercial platform, like those developed by Parscale. And never assume that your algorithms are as good as you believe. Instead test and refine them constantly.

As the meeting concluded I asked a question similar to the one I had put to the Democrats. "Did you win because you had the best data operation or despite the fact the Clinton campaign had a better operation?"

Their response was as swift as the answer we had heard from the Democrats earlier in the day. "No question we had the better data operation. We saw Michigan start to break for Trump before the

Clinton campaign did. And we saw something else that the Clinton team never saw. We saw Wisconsin break for Trump the weekend before election day."

After both political teams had left, I turned to the Microsoft team and asked for a show of hands. Who thought the Clinton team had the better data operation and who thought the RNC/Trump team had the better operation? The vote was unanimous. Everyone in the room concluded that the approach used by Reince Priebus and the Trump campaign was superior.

The Clinton campaign had relied upon its technical prowess and its early lead. The Trump campaign, by contrast and out of necessity, had relied on something closer to the shared-data approach Matthew Trunnell described.

There will always remain plenty of room to debate the various factors that determined the outcome of the 2016 presidential race, especially in battleground states where the votes were close, such as Michigan, Wisconsin, and Pennsylvania. But as we concluded that day, Reince Priebus and the RNC's data model quite possibly helped change the course of American history.

If a more open approach to data can do that, just imagine what else it could do.

The key to this type of technology collaboration lies with human values and processes and not just a focus on technology. Organizations need to decide whether and how to share data, and if so, on what terms. A few principles will be foundational.

The first is concrete arrangements to protect privacy. Given the evolution of privacy concerns, this is a prerequisite both for enabling organizations to share data about people and for people to be comfortable sharing data about themselves. A key challenge will involve the development and selection of techniques to share data while protecting privacy. This will likely include new so-called "differential privacy" techniques that protect privacy in new ways, as well as providing access to aggregated or de-identified data or enabling query-only access to a data set. It may also involve the use of machine learning that is trained on encrypted data. We may well see new models emerge that enable people to decide whether to share their data collectively for this type of purpose.

A second critical need will involve security. Clearly, if data is federated and accessible by more than one organization, the cybersecurity challenges of recent years take on an added dimension. While part of this will require continuing security enhancements, we'll also need improvements in operational security that enable multiple organizations to manage security together.

We'll also need practical arrangements to address fundamental questions around data ownership. We need to enable groups to share data without giving up their ownership and ongoing control of the data they share. Just as landowners sometimes enter into easements or other arrangements that allow others onto their property without losing their ownership rights, we'll need to create new approaches to manage access to data. These must enable groups to choose collaboratively the terms on which they want to share data, including how the data can be used.

In addressing all these issues, the open-data movement can take a page from open-source trends for software. At first that effort was hampered by questions about license rights. But over time standard open-source licenses emerged. We can expect similar efforts for data.

Government policies can also help advance an open-data movement. This can begin by making more government data available for public use, thereby reducing the data deficit for smaller organizations. A good example was the decision by the US Congress in 2014 to pass the Digital Accountability and Transparency Act, which makes publicly available more budget information in a standardized way. The Obama administration built on this in 2016 with a call for open data for AI, and the Trump administration has followed by proposing an integrated federal data strategy to "leverage data as a strategic asset" for government agencies.[15] The United Kingdom and the European Union are pursuing similar efforts. But today only one in five government data sets is open. Much more needs to be done.[16]

Open data also raises important issues for the evolution of privacy laws. Current laws were mostly written before AI developments began to accelerate, and there are tensions between current laws and open data that deserve serious consideration. For example, European privacy laws focus on so-called purpose limitations that restrict the use of information only for the purpose specified when data was collected. But many times new opportunities emerge to share data in ways that will advance societal goals—like curing cancer. Fortunately, this law allows data to be repurposed when it is fair and compatible with the original purpose. Now there will be critical questions about how to interpret this provision.

There will also be important intellectual-property issues, especially in the copyright space. It has long been accepted that anyone can learn from a copyrighted work, like reading a book. But some now question whether this rule should apply when the learning is conducted by machines. If we want to encourage broader use of data, it will be critical that machines can do so.

After developing practical arrangements for data owners and addressing government policy, one more need will be vital. This is the development of technology platforms and tools to enable easier and less costly data sharing.

This is one of the needs that Trunnell has encountered at the Hutch. He's taken note of the difference between the work pursued by the cancer research community and by tech companies. New, cutting-edge tools for managing, integrating, and analyzing diverse data sets are being developed by the technology sector. But as Trunnell recognized, "the divide between those producing the data and those building novel tools is a huge missed opportunity for making impactful, life-changing—and potentially lifesaving—discoveries using the massive amount of scientific, educational, and clinical trial data being generated every day."[17]

But for this to be viable, data users need a strong technology platform that is optimized for open-data use. Here the market is starting to go to work. As different tech companies consider different business models, they have alternatives from which to choose. Some may choose to collect and consolidate data on their own platform and offer access to their insights as a technology or consulting service. In many respects, this is what IBM has done with Watson and what Facebook and Google have done in the world of online advertising.

Interestingly, as I was listening to Matthew Trunnell that August evening, a team from Microsoft, SAP, and Adobe was already working on a different but complementary effort. The three companies announced the Open Data Initiative, launched a month later, designed to provide a technology platform and tools to enable organizations to federate data while continuing to own and maintain control of the data they share. It will include tech tools that organizations can use to identify and assess the useful data they already possess and put it into a machine-readable and structured format suitable for sharing.

Perhaps as much as anything else, an open-data revolution will require experimentation to get this right. Before our dinner ended, I pulled up a chair next to Trunnell and asked what we might do together. I was especially intrigued by the opportunity to advance work that we at Microsoft were already pursuing with other cancer institutes in our corner of North America, including with leading organizations in Vancouver, British Columbia.

By December, this work had borne fruit and we announced a

$4 million Microsoft commitment to support the Hutch's project. Formally called the Cascadia Data Discovery Initiative, the work is designed to help identify and facilitate the sharing of data in privacy-protected ways among the Hutch, the University of Washington, and the University of British Columbia and the BC Cancer Agency, both based in Vancouver. It is an early example of what is starting to spread, including the California Data Collaborative, where cities, water retailers, and land planning agencies are federating data to enable analytics-driven solutions to address water shortages.[18]

All of this provides cause for optimism about the future of open data, at least if we seize the moment. While some technologies benefit some companies and countries more than others, that is not always the case. For example, nations have never been forced to grapple with hard questions about who would be the world leader in electricity. Any country could put the invention to use, and the question was who would have the foresight to apply it as broadly as possible.

Societally we should aim to make the effective use of data as accessible as electricity. It is not an easy task. But with the right approach to sharing data and the right support from governments, it is more than possible for the world to create a model that will ensure that data does not become the province of a few large companies and countries. Instead it can become what the world needs it to be—an important engine everywhere for a new generation of economic growth.

Chapter 16 ⟫ **CONCLUSION:**
Managing Technology That
Is Bigger Than Ourselves

When Anne Taylor was a teenager at the Kentucky School for the Blind, she tapped into a passion that flourished into a career. By day, Anne helps make our products more accessible to people with disabilities. She says she loves her job. But Anne lights up even more when she talks about how she spends her free time. "I'm a hacker," she says.

In 2016, Anne was the second hacker recruited for a project using AI, computer vision, and a smartphone camera. One of her tasks was testing the app by walking around Microsoft's campus with a phone strapped to her forehead. The life of an inventor is sometimes not for the fashion conscious. But when it comes to fashion, almost anything seems to go at Microsoft.

The team's work led to a breakthrough, an AI-powered app that helps people who are blind "see" the world as it's described through their smartphones. With Seeing AI, Anne, who is blind herself, can now read a handwritten note from her family, independently. As she says, "That probably seems simple to you guys, since you've been able to do this for a long time. But when someone writes something to me that's personal or private in nature, I've always had to ask someone else to read it to me. Now I don't. That means something."[1]

Recognizing text isn't important just for reading a modern-day letter. In New Jersey, AI has altered the research of Marina Rustow, a professor of Near Eastern studies at Princeton University's Geniza Lab, where she interprets and decodes a massive trove of four hundred thousand documents from Cairo's Ben Ezra Synagogue, the largest recorded cache of Jewish manuscripts.

Studying these documents is a formidable challenge. Many are torn into fragments and scattered in libraries and museums around the world. The sheer volume and location of the material makes physically piecing them together next to impossible. With AI, Rustow's team was able to comb through the digital fragments and match pieces stored thousands of miles apart, painting a previously incomplete picture of how Jews and Muslims coexisted in the middle ages.[2]

If an AI algorithm can help Rustow preserve the distant past, what can it do to protect the world's living history?

In Africa, poaching is an ongoing issue that threatens to snuff out endangered species, including some of the world's most iconic and recognized animals. Microsoft's AI for Earth team is working with researchers at Carnegie Mellon University to help park rangers in the Uganda Wildlife Authority stay one step ahead of poachers. Using an algorithm to sift through fourteen years of historical national park patrol data, the Protection Assistant for Wildlife Security application, or PAWS, uses computational game theory to learn and predict poaching behavior, enabling authorities to proactively identify poaching hotspots and modify their patrols.[3]

As these examples show us, the power of technology can help the blind see the world in new ways, historians discover the past, and scientists pursue new strategies for an ailing planet. Its potential promise has virtually unlimited breadth.

AI is unlike singular inventions from the past such as the automobile, the telephone, or even the personal computer. It behaves more like electricity in that it powers tools and devices that run almost every aspect of society and our lives. Like electricity, AI will run in the background and in many ways, we will forget it's even there, until the day the power goes out.

Satya has dubbed this new reality "tech intensity," a term that describes the infusion of technology into the world around us.[4] This new era is an opportunity for companies, organizations, and even entire countries to spur their growth by not just adopting technology, but building their own, which creates an imperative for organizations to equip their employees with the new skills and capabilities needed to put this technology to work.

It's a time of enormous promise, but also of new challenges. Digital technologies literally have become both tools and weapons. They take us back to Albert Einstein's words in 1932, reminding people of the benefits created by the machine age but calling on humanity to ensure that its organizing power keeps pace with its technical advances.[5] As we keep working to bring more technology to humanity, we also need to bring more humanity into technology.

As described in these chapters, technology today is having an immensely uneven economic impact, creating huge advances and wealth for some while leaving others behind as it displaces jobs and fails to reach communities that lack broadband connectivity. It's changing the face of war and peace, creating a new theater of warfare in cyberspace and new threats to democracy through state-sponsored attacks and disinformation. And it's increasing the polarization of domestic communities, eroding privacy, and creating an emerging capability for authoritarian regimes to exercise unprecedented surveillance of their citizens. As AI continues to advance, all these developments will accelerate even more.

We see these dynamics come into play in the political issues of our time. People argue about immigration, trade, and tax rates for wealthy individuals and corporations, but we seldom see politicians consider or the tech sector acknowledge the role that technology is playing in creating these challenges. It's as if we're all so absorbed in the resulting symptoms that we lack the time and energy to focus on some of the important underlying causes. Especially as the impact of technology continues to accelerate, it risks fostering a myopic understanding.

It's unrealistic to expect the pace of technological change to slow. But it's not too much to ask that we do more to manage this change. In contrast to prior technology eras and inventions such as the railroad, telephone, automobile, and television, digital technology has progressed for several decades with remarkably little regulation—or even self-regulation. It's time to recognize that this hands-off attitude needs to give way to a more activist approach that addresses evolving challenges in a more assertive way.

A more active approach doesn't mean that everything should be left to governments and regulation. That would be as shortsighted and unsuccessful as asking governments to do nothing at all. To the contrary, this needs to start with individual companies and with more collaborative work across the tech sector.

When Microsoft was in the hot seat two decades ago, we recognized that we needed to change. I took from our battles three lessons that we continue to learn from and apply. As we consider the current role of technology in the world, these seem as applicable to the entire tech sector today as they did to our company in the past.

First, we needed to accept the heightened expectations that those in government, the industry, our customers, and society at large had for us. We had to assume more responsibility, whether it was required by law or not. We were no longer an upstart. We needed to strive to set an example rather than argue that we could do whatever we wanted.

Second, we needed to get out and listen to what other people had to say and do more to help solve the technology problems that needed to be solved. This started with building constructive working relationships with more people. But that was just the start. We had to understand better other people's perceptions of and concerns about us. We needed to do a better job of solving small problems before they grew out of control. This required that we sit down more frequently with governments and even our competitors to find common ground. We recognized that we'd undoubtedly face some hard questions, and we would need to find the courage to compromise.

There were days when some engineers argued that we should instead keep on fighting. At times, I almost felt that they were calling my courage into question. While there were times when we needed to stand our ground, there were many moments when I argued that it took more bravery to compromise than it did to keep fighting. And it took persistence as well. The quest for common ground often led to negotiations that ended in impasse and failure before we could come back together and reach an agreement. We needed to develop the ability to fail gracefully, complimenting the other side even when things fell apart so we could preserve the ability to work through the hard problems again when the right moment arrived. It almost always did.

And finally, we needed to develop a more principled approach to our work. We needed to maintain an entrepreneurial culture, while also integrating this with principles that we could talk about both internally and externally. We began to develop the capability to craft such principles, first for antitrust issues and later for interoperability and human rights questions. As Satya suggested in 2015 for surveillance issues, as discussed in chapter two, we developed principles to guide our decision making for this topic as well. The resulting cloud commitments continue to serve as a model for us in other areas. Among many other virtues, this approach both helps and forces us to think about the responsibilities we bear and the best ways to address them.

In many ways, these approaches call for a cultural change across the tech sector. For a lot of good reasons, tech companies have traditionally focused first on developing a product or service that is exciting and then on attracting as many users as possible as quickly as possible. There has often been little time or attention beyond this. As Reid Hoffman has captured accurately in his term *blitzscaling*, a "lightning-fast path" that prioritizes speed over efficiency provides the best approach to developing market-leading technology on a global scale.[6] Even when companies achieve this type of leadership position, there remains an ongoing need to move quickly. It's easy to imagine the concerns that would arise in Silicon Valley when weighty demands threatened to slow down innovation.

These concerns are important. But given the role that technology now plays in the world, it's equally dangerous for a tech company to move faster than the speed of thought, or simply to fail to think at all about the broader implications of its services or products. One thesis of this book is that it is more than possible for companies to succeed while doing more to address their societal responsibilities. As Satya is quick to point out when such issues arise, we need to move fast but with some guardrails on our technology. An ability to anticipate issues and define a principled approach to address them is more likely to keep "the car on the road" as it gains speed. It helps avoid at least some of the public controversies and potential reputational damage that can force executives to spend more time on issues other than product development and user growth.

But even with the best of intentions, this type of effort does not come easily. The most natural course is to keep expanding a product and sell it to anyone who will buy it. There almost always arise internal objections when there is a discussion about self-restraint. (I speak from experience.) A commitment to regulate a company's own conduct therefore requires leadership from the top. Senior leaders need to think broadly, and they need to encourage their people to do more than simply find problems with every potential solution and instead help find solutions for every potential problem.

Part of the answer requires that tech companies develop greater capabilities in disciplines beyond traditional product development, marketing, and sales. As technology collides with the issues of the world, there is no substitute for strong leaders in finance, legal, and human resources. In the past, these roles in the tech sector were sometimes thought of as important principally for raising capital, selling a company, or going public. Today's issues and needs are far broader.

One reason these disciplines are important is that it's not easy to define overarching principles to govern a product's path. It requires considered thought and a strong understanding of public expectations, real-world scenarios, and practical development needs, all of which depend on close interaction with engineering and sales teams. During a typical week at Microsoft, it's not unusual to find Dev Stahlkopf, our general counsel, spending part of her time working with people on projects that seek to look around corners to get ahead of looming problems and controversies.

Another challenge is that this work isn't finished when new principles are adopted. As our own internal audit team advised Amy

Hood and me, it was critical that we build on our identification of the ethical issues for AI discussed in chapter eleven not just by defining new principles, but by establishing concrete policies, a governance structure, an accountability framework, and employee training to make this ethical commitment effective. This points to one of the biggest tests for large and established tech companies that serve hundreds of millions of customers around the world. Principles need to be operationalized and implemented on a global scale, as our engineering work to implement the GDPR illustrates in chapter eight. It's the type of work that requires broad support from all the various disciplines that contribute to the running of a contemporary global corporation.[7]

Ultimately, well-informed and broad-minded leadership needs to translate into more proactive steps at individual tech companies and also more collaboration across the tech sector as a whole. Compared to many other industries, the tech sector today is often fragmented and even divided when it comes to trade associations and voluntary efforts. Given the diversified nature of technology and competing business models, this isn't altogether surprising. But even with ongoing differences, there is room for the tech sector to do more together.

This need is especially pronounced when it comes to priorities such as strengthening cybersecurity and combating disinformation. There have been important recent initiatives, like the response to WannaCry described in chapter four and Siemens's Charter of Trust and the broader Cybersecurity Tech Accord described in chapter seven. But in some ways, these have just scratched the surface in terms of what is possible and what the public and governments increasingly will expect.

This too will require cultural change. Too often today, even leading tech companies find it easy to look at a problem like cybersecurity and conclude that they don't need to work closely with the rest of the industry. Or they decide they won't join a parade unless they get to lead it. Or they decide that they won't be part of a parade if it includes some other tech company that's currently on the hot seat politically, for fear of being "infected" by standing near someone who is facing public criticism. While these concerns to some degree are understandable, it's important for tech leaders to resist them. Collectively such views make it more difficult for the tech sector to address the responsibilities that the world expects of it.

While there is enormous opportunity for individual companies and the tech sector to collectively do more, it's impossible to conclude that this can relieve governments of their responsibility to do more as well. The tech sector is full of good and thoughtful people, but the three centuries since the dawn of the industrial revolution are devoid of any major industry successfully regulating everything about itself completely by itself. It would be naive to think the first successful case will emerge now.

Even if it is possible, we should question whether this would be the best path forward. The sweep of technology issues impacts virtually every aspect of our economies, societies, and personal lives. In the democracies of the world, one of our most cherished values is that the public determines its course by electing the people who make the laws that govern us all. Tech leaders may be chosen by boards of directors selected by shareholders, but they are not chosen by the public. Democratic countries should not cede the future to leaders the public did not elect.

All of this makes it important for governments to take a more active and assertive approach to regulating digital technology. Like everything else in this book, this is easier said than done. But there are some important lessons worth applying.

For one, there's a strong case for governments to innovate in the regulatory space in a way that's like innovation in the tech sector it-self. Instead of waiting for every issue to mature, governments can act more quickly and incrementally with limited initial regulatory steps—and then learn and take stock from the resulting experience. In other words, take the concept of a "minimum viable product" and consider the type of approach we advocated for AI and facial recognition, described in chapter twelve. We readily recognized that just as for a new business or software product, the first regulatory step would not be the last, but we believed it was wiser for governments to take a series of more limited steps more quickly.

Is this an approach that can work in certain areas for technology regulation? If so, it could become a new regulatory tool for our time. If governments can adopt limited rules, learn from the experience, and subsequently use this learning to add new regulatory provisions much as companies add new features to products, it could put laws on a path to move faster. To be clear, officials must still consider broad input, remain thoughtful, and be confident that they have the right answers for at least a limited set of important questions. But by bringing some of the cultural norms developed in the tech sector into the regulation of technology itself, governments can do more to catch up with the pace of technological change.

Governments can also have a more positive and practical impact if they do more to take stock of changing technology trends and look for opportunities to stimulate market solutions more broadly. Our approach to rural broadband, as described in chapter nine, is based on this concept. Rather than expensive public investments in costly fiber-optic cables that will take decades to reach rural homes, it's more sensible for government funding to stimulate new wireless technologies in ways that will accelerate market forces so they can reach a critical liftoff point and then move forward on their own.

More than ever, governments have multiple opportunities to stimulate technology market forces through their actions. They typically rank among a nation's largest technology purchasers, and their procurement decisions can have a powerful impact on overall market trends. Even more important, governments have large and valuable data repositories. By making this data available for public use in appropriate and defined ways, governments can have a decisive influence on the technology markets that will put this data to use. For example, they can help stimulate better-informed public-sector and civic efforts to match the skills needed for new jobs with people who want to pursue them, as described in chapter ten. And it provides a powerful tool that governments can use to help accelerate the adoption of open-data models, as discussed in chapter fifteen.

A more active regulatory approach will require that government officials develop an even greater understanding of technology trends. This in turn will require more conversation between those who create technology and those who must regulate it. This too has more than its share of challenges. Historically there has never been a national business or technology center as distant from a country's capital as Silicon Valley is from Washington, DC. Even this fails to capture the distance between America's political and technology capitals. As Margaret O'Mara, a historian at the University of Washington, noted, "Operating far away from the centers of political and financial power in a pleasant and sleepy corner of Northern California, they created an entrepreneurial Galapagos, home to new species of companies, distinctive strains of company culture, and tolerance for a certain amount of weirdness."[8]

A geographic gulf of almost twenty-five hundred miles obscures one thing the two places have in common. Traveling to each from a place like Seattle (which has its own tolerant appreciation for weirdness), you can understand, given the excitement and activity in each location, why it's easy to feel, once there, that each place is at the center of the world. But more than ever, there's a need to build a stronger bridge across this geographic divide.

One challenge is that many in tech circles for too long have asserted that people in government don't understand enough about technology to regulate it properly—even while tech companies benefited from all manner of government funding and support.[9] It's a view the press is too quick to reinforce as it jumps on the mistakes that legislators sometimes make if they ask a tech executive the wrong question or even the right question the wrong way. But in my experience, government officials have come a long way since the morning fifteen years ago when I was talking about digital advertising with a US Senator who was unaware that he could read the *Washington Post* on the internet.

Having worked in the technology sector for more than a quarter century, I realize that the products are complex. But so are contemporary commercial airplanes, automobiles, skyscrapers, pharmaceuticals, and even food products. You don't hear any serious suggestion that the Federal Aviation Administration should leave aircraft unregulated because they are too complicated for people in government to understand.[10] The flying public would not stand for it. Why is information technology fundamentally different, especially when many of an airplane's components are now based on it?

The truth is that government agencies have long proven adept at developing the factual capacity to understand the products they regulate. This doesn't mean the process is free of frustration or that everyone does an equally good job. Nor does it mean that all regulatory approaches make good or even common sense. But the tech sector needs to get over any illusion that it alone is capable of understanding information technology and its intricacies. Instead, it will need to do more to share information about these nuances so the public and governments can better appreciate them.

In many ways, a second challenge for governments is far more pronounced. Information technology and the companies that create it have gone global. The internet was designed to be a global network, and many of its benefits come from its connected nature. Perhaps more than any other technology in history, its influence and its geographical reach exceed any single government. This sets it apart from prior inventions such as the telephone, television, and electricity, which are based on networks or grids that typically correspond to national or state lines.

One way to appreciate this challenge is to consider the technology that perhaps was the most similar to digital technology in its regulatory impact. As the 1800s progressed, railroads played arguably a bigger role than any other invention in redefining the United States. They extended beyond the boundaries of the state governments that initially had assumed the most authority to regulate the economy. In the decades that followed the Civil War, the nation's railroad companies in many ways became larger and more powerful than many state governments.

Things came to a head in the 1880s. There was virtually no tradition of regulating the economy at the federal level except during times of war, and proposals in Washington, DC, to regulate railroads, repeatedly went down in defeat. The state governments responded by passing laws to regulate railroad rates that impacted trips beyond their borders. In 1886, the Supreme Court struck these down, ruling that the federal government alone had this power.[11] Suddenly the public confronted a stark reality: The states "could not, and the federal government would not, regulate railroads."[12] This new political dynamic broke the impasse, and the next year Congress created the Interstate Commerce Commission to regulate railroads.[13] The modern federal government was born.

The global reach of contemporary information technology is akin

to the railroad tracks of the 1880s that kept progressing beyond juris-
dictional lines. But today there is no global counterpart to the Interstate Commerce Commission. And understandably, there exists no
appetite for creating one.

How can governments regulate a technology that is bigger than
themselves? This is perhaps the single greatest conundrum confronting technology's regulatory future. But once you ask the question,
one part of the answer becomes clear: Governments will need to
work together.

There exist many hurdles that will need to be overcome. We live
in a time when roiling geopolitical headwinds are causing many governments to pull inward. It's difficult to expect great leaps in bringing nations together when the day's dominant headlines talk about
countries leaving trade blocks or pulling out of long-standing treaties.
Beyond this, it's a time when many governments are finding it difficult even to make decisions that matter only to themselves.

But amid these pressures, the inexorable course of technology is
forcing more international collaboration. As this book has illustrated,
issues like surveillance reform, privacy protection, and cybersecurity
safeguards have all required governments to deal with each other in
new ways. It's one reason so many of our initiatives at Microsoft have
focused on supporting the building blocks needed for international
progress. Since the start of 2016, these include the coordinated response to WannaCry, the industry's Cybersecurity Tech Accord, the
multi-stakeholder Paris and Christchurch calls, the US-EU Privacy
Shield, the CLOUD Act's authorization for international agreements,
and a long-term vision for a Digital Geneva Convention. These same
years have seen stronger privacy protection move across the Atlantic
and the emergence of a new global conversation about AI and ethics.
If this type of progress is possible in a time of growing nationalism,

there is hope for even more headway when the international pendulum swings back toward the center.

To start with, we will need to continue to build coalitions of the willing. Six governments and two companies came together publicly to address WannaCry. A group of thirty-four companies launched the tech accord, and an initial group of fifty-one governments were part of the multi-stakeholder support for the Paris Call. In each instance, there were important and even critical omissions. But progress came not by dwelling on who was missing but on who could be persuaded to join. This in turn led to continued momentum and additional expansion later on.

We also need to recognize that some issues may lead to global consensus and others may not. Many of today's technology issues involve questions of privacy, free expression, and human rights that lack global support. A coalition of the willing is most likely to require the world's democratic countries to come together. This is not a small group. Today there are roughly seventy-five democratic nations with a total population approaching four billion.[14] This means that more people live in democratic societies than at any time in history. But recently the world's democracies have become less healthy. Perhaps more than that of any group of societies, their long-term well-being requires new collaboration to manage technology and its impact.

This makes it even more important to sustain momentum until the day the United States government resumes its long-standing diplomatic role by both supporting and providing leadership for these types of multilateral initiatives. There is no mistaking the fact that the world's democracies are weaker when the United States is standing apart from the rest.

Continued progress also requires that governments recognize that in addition to regulating technology, they need to regulate themselves.

Issues like cybersecurity and disinformation will shape the future of war and the protection of our democratic processes. Just as no industry in history has fully engaged in successful self-regulation on its own, there's no precedent for a nation protecting itself by relying solely on the private sector or even by regulating it. Governments will need to act together, and part of this will require new international norms and rules that limit national conduct and hold countries accountable when they violate those rules.

This inevitably will lead to new debates about the virtues of international rules. You can already hear the concerns that will be voiced regarding the probability that some countries will follow these rules while others will not. The world has had arms control bans and limitations in place since the latter 1800s, and for over a century, controversies have always swirled around the same points. The harsh reality is that some countries will violate these agreements. But it is easier for the rest of the world to respond effectively when an international norm or rule is in place.

Digital technologies' new challenges also require more active collaboration across traditional institutional boundaries. You can see this, for example, in successful projects to help manage technology's broad societal impact by bringing together governments, nonprofit groups, and companies to address jobs and the need for people to develop new skills, as described in chapters ten and thirteen. This type of combination can also help address other community challenges such as affordable housing, as reflected in recent initiatives in the Seattle area.

But the opportunity and need for new forms of such collaboration does not end with these social issues. More than ever before, the protection of fundamental human rights rests on steps that governments, NGOs, and companies must take together. This will continue to become more pronounced as even more data moves to the cloud and more governments push for data centers to be constructed within their borders. The issues of the twenty-first century require initiatives that are both multilateral and multi-stakeholder in scope.

One key to multi-stakeholder collaboration is the recognition of the respective roles that each group needs to play. Government officials play a unique leadership role, especially in democratic societies, as those elected by the people to make societal decisions. They alone have the authority and responsibility to chart the course of public education and make and apply the laws under which we all live. Companies and nonprofit groups can bring a civic spirit and complement and partner with governments, bringing additional resources, expertise, or data the public sector often needs. And companies and nongovernmental organizations can test new ideas by experimenting and moving faster, especially across borders. We all need to appreciate and respect each other's roles.

Many issues also will require compromise. For successful business leaders who have helped build some of the world's most valuable companies, this is not always easy to contemplate. They typically succeeded against long odds by doing things their own way; regulation will restrain their freedom in the future.

This perhaps explains why some tech leaders argue in public and assert even more in private that the greatest risk to innovation is that governments will overreact and overregulate technology. It's a clear risk, but we currently remain far from falling over this precipice. Politicians and officials have started calling for regulation, but so far there's been a lot more talk than action. Rather than fret excessively about the dangers of overregulation, the tech sector would be better served by thinking about what shape intelligent regulation should take.

There's a final consideration, and it's the most important. These issues are bigger than any single person, company, industry, or even technology itself. They involve fundamental values of democratic freedoms and human rights. The tech sector was born and has grown because it has benefited from these freedoms. We owe it to the future to help ensure that these values survive and even flourish long after we and our products have passed from the scene.

This context provides clarity. The greatest risk is not that the world will do too much to solve these problems. It's that the world will do too little. And it's not that governments will move too fast. It's that they will be too slow.

Technology innovation is not going to slow down. The work to manage it needs to speed up.

Acknowledgments

Neither of us has written a book before, so we leaned on many people to help us through this process. One thing we learned is hardly surprising: There is a big difference between reading a book and writing one. While reading a good book is an adventure, writing one is like embarking on the Iliad.

Our journey began in a booth at a trattoria in New York's Gramercy Park, where we met Tina Bennett of William Morris Endeavor. As neophytes to the publishing world, we were a little surprised, but more than delighted, when she agreed to work with us as our literary agent. And work with us she has! She has not only helped map our journey but traveled with us on it. Tina helped us find the book within us, navigate each step, and advised us as we refined chapters, pages, and even words. We've also benefited from working closely with Tina's colleagues at WME, Laura Bonner and Tracy Fisher, who guided us through the international publishing process.

The single biggest step Tina helped us take was to the doors of Penguin Press. We knew Scott Moyers was the right editor as soon as we sat in his office. We held our breath as we waited for publishers' bids to arrive and exhaled when it was clear Scott's enthusiasm was mutual. Since that day, Scott has provided gentle but clear feedback, and we've heeded his words. He and assistant editor, Mia Council, have turned around material at a speed that's as fast as anything we've seen in the tech sector, which was vital to maintaining our hectic schedules. When the writing and editing was complete, Penguin's crack marketing team, including Colleen Boyle, Matthew Boyd, Sarah Hutson, and Caitlin O'Shaughnessy, took the book to market. It has been a delightful partnership from beginning to end.

This project wouldn't have been possible without the support of several people at Microsoft, starting with Satya Nadella. As an author himself, he appreciated the opportunity the book offered both to think more deeply and engage more broadly on the issues technology is creating for the world.

And he read and provided feedback as our work progressed. Frank Shaw brought his keen eye and good judgment to our draft, as he does for every aspect of our public communications. And Amy Hood, as always, shared her keen intellect and practical wisdom while providing both moral support and a little laughter from down the hall.

As we advanced, we benefited enormously from individuals who gave generously of their time, providing a bit more distance from and a broader perspective on our efforts. The first was Karen Hughes, who took her varied public communications expertise and applied it to an early draft. Our dinner with her in Washington, DC, was not just a detailed editorial review but a penetrating communications workshop. It was yet another reminder of why we turn to her for advice when we need to navigate big communications challenges.

As we approached a complete draft, huge help came from David and Katherine Bradley and their son Carter, who generously gave us their time to read and share detailed reactions, both in person and with their written notes. Their thoughtful and multigenerational feedback improved our book in multiple places.

As we neared the finish line, David Pressman read the entire manuscript thoroughly and offered both an assessment of some weaknesses and constructive suggestions to address them. He provided an experienced diplomat's perspective on the human rights and international relations challenges that increasingly have defined today's tech sector.

Throughout the writing and editing process, a few people played outsized roles in helping us with research and fact-checking, including Jesse Meredith, whom we met as a post-doctorate history student at the University of Washington and is now teaching at Colby College in Maine. Stephanie Cunningham, a librarian at the Microsoft Library, provided blazingly fast and accurate answers even to our most obscure questions. The Library is a vital resource on our Redmond campus. And what would we have done without Maddie Orser, who dusted off her master's degree in history to fact-check our historical references and help perfect our endnotes? A special thanks also goes to Microsoft's Thanh Tan, who not only has a bloodhound's nose for a good story, but can find the right person to tell it and get invited to their home for dinner.

We also relied heavily on Dominic Carr, who was critical to the initial conception of the book, helpful as we took each step, and vital as we approach the broader public conversations we hope the book will encourage. As we juggled our work and writing, we were backed by the small team that shares our corridor on Microsoft's Redmond campus—most notably, Kate Behncken, Anna Fine, Liz Wan, Mikel Espeland, Simon Liepold, Katie

Bates, and Kelsey Knowles. We also relied on Matt Penarczyk as the lawyer who negotiated the publishing contracts on behalf of Microsoft.

In the final stages, we turned to many additional colleagues and friends to review and check our facts. Inside Microsoft this included Eric Horvitz, Nat Friedman, Harry Shum, Fred Humphries, Julie Brill, Christian Belady, Dave Heiner, David Howard, Jon Palmer, John Frank, Jane Broom, Hossein Nowbar, Rich Sauer, Shelley McKinley, Paul Garnett, Dev Stahlkopf, Liz Wan, Dominic Carr, Lisa Tanzi, Tyler Fuller, Amy Hogan-Burney, Ginny Badanes, Dave Leichtman, Dirk Bornemann, and Tanja Boehm. Hadi Partovi and Naria Santa-Lucia checked the accuracy of our writing regarding their organizations. Jim Garland and his team at Covington & Burling then added a careful legal review of certain sensitive issues, as did Nate Jones, from his new consulting firm.

A special thanks goes to Microsoft graphic designers Mary Feil-Jacobs and Zach LaMance, for moonlighting on the book's cover art.

We're also indebted enormously to the many colleagues, peers, and friends inside and outside Microsoft who played vital roles in the events that are captured in the book.

This starts with the extraordinary trio of Bill Gates, Steve Ballmer, and Satya Nadella, the three individuals who have served as CEO of a company that has had a truly remarkable history. Very few people have had the opportunity to work closely with all three. Each is different from the other, but they share the broad curiosity and passionate pursuit of excellence that it takes to truly make a difference in the world. And that's just for starters.

Especially important as well are the members of Microsoft's Senior Leadership Team and Board of Directors, and the members of the Senior Leadership Team of the company's Department of Corporate, External, and Legal Affairs. In so many ways, they represent the tiny tip of an enormous iceberg of people whom we're so lucky to know. The opportunity to contribute to world-changing technology brought each of us into the tech sector. But the chance to work with wonderful people and forge lasting friendships is a big part of what keeps us here.

We also want to thank the many other people we have the continuing opportunity to work with, including at other tech companies, among governments around the world, across the nonprofit sector, and with many journalists around the world. Hopefully, you'll find our references to you at least reasonably fair. That was our goal. At the end of the day, we each often approach these issues from different perspectives, but it's our collective ability to formulate a common understanding that will shape technology's connection with the world.

We also would be remiss without recognizing the individuals and groups that work at Microsoft under Rajesh Jha who create the tools that enabled us to be so productive and efficient. For a book like this, Microsoft Word remains an author's best friend. It's perhaps easy for people on some days to take its widely varied features for granted. We most definitely did not, whether it was for the formatting of hundreds of endnotes or the use of Word Online that enabled us to write and edit at the same time, on the same manuscript from different locations. Other products like OneNote and Teams helped us collaborate on research, interviews, and notes, and OneDrive and SharePoint helped us organize, store, and share all of our work. One of our favorite tools was one of the company's newer products, the To-Do app, which we used to create shared lists tracking the project's many tasks.

Over the course of the year that we worked on this book, our "day jobs" took us to meetings, events, and public presentations in twenty-two countries on six continents, as well as to numerous places across the United States. It all helped shape our thinking, and many of these experiences are reflected in the stories we share in *Tools and Weapons*. But it also meant that the writing, especially over the course of six intense months, took place for both of us during many early mornings, late nights, on weekends, and even on vacations and holidays.

All this required a lot from our families, and to them we owe our greatest gratitude. They've always provided love and support even when work has involved a global travel schedule or interruptions on the weekends. And this book required even more of their help. Our respective spouses—Kathy Surace-Smith and Kevin Browne—read perhaps more chapters than they wanted, offered helpful insights and suggestions, and supported the whole project with the patience of saints. We each have two children, and in important ways the book became a bit of a family affair. At times, our two families came together so life could continue apace with our work, and we'll always smile as we think about editing during the day and playing board games with our families in the evening.

As all this reflects, this was both a happy adventure and an arduous journey. As it reaches a conclusion, we want to thank everyone who made it possible.

Brad Smith
Carol Ann Browne
Bellevue, Washington

Notes

INTRODUCTION: THE CLOUD

1. The earliest archives contained data that would be at home in a modern data center. For example, archeologists have discovered at the site of ancient Ebla in Syria the remains of a royal archive that was destroyed around 2300 BC. In addition to the text of a Sumerian myth and other documents used by palace scribes, there were two thousand clay tablets filled with administrative records. These contained details about the distribution of textiles and metals, as well as cereals, olive oil, land, and animals. Lionel Casson, *Libraries in the Ancient World* (New Haven, CT: Yale University Press, 2001), 3–4. It's easy to imagine a contemporary data analytics team working with similar data sets in our own day.

   In the ensuing centuries, libraries spread around the ancient Mediterranean to the thriving city states of Greece, then to Alexandria, and ultimately to Rome. Their collections diversified as humanity discovered its voice and improved its ability to store written works on papyrus scrolls rather than clay tablets. The main library in Alexandria, founded around 300 BC, had 490,000 rolls of documents. Casson, *Libraries*, 36. At the same time, private libraries sprang up in east Asia, with collections stored in bamboo chests. The invention of paper in China in 121 AD represented a major breakthrough and "put the East centuries ahead of the West, enabling elaborate systems of administration and bureaucracy to be created." James W. P. Campbell, *The Library: A World History* (Chicago: The University of Chicago Press, 2013), 95.

2. The story of the invention of the filing cabinet illustrates the changing needs over time for the storage of data. In 1898, Edwin Siebel, an American insurance agent, grew frustrated with the data storage techniques of the time. Siebel lived in South Carolina and insured cotton as it made its way from the fields, over the Atlantic, to textile mills in Europe. It required substantial paperwork that needed safekeeping. In Siebel's time, businesses filed their records in wooden "pigeonholes" stacked floor to ceiling along the walls. Papers were typically folded, tucked in envelopes and slid into cubbies, often requiring a ladder to reach them. It wasn't an easy or efficient way to store information, especially when someone had to hunt down a document and wasn't sure where it was stored.

   Like any good inventor, Siebel saw a problem that needed to be solved. He developed a simple yet clever idea: a vertical filing system stored in a wooden box. He worked with a manufacturer in Cincinnati to build five boxes with drawers that stored papers on end, allowing a clerk to quickly flip through and read files without opening a single envelope. Eventually, these papers would hang in folders, separated by labeled tabs sandwiched between. The modern filing cabinet was born. James Ward, *The Perfection of the Paper Clip: Curious Tales of Invention, Accidental Genius, and Stationery Obsession* (New York: Atria Books, 2015), 255–56.

3. David Reinsel, John Gantz, and John Rydning, *Data Age 2025: The Digitization of the World From Edge to Core* (IDC White Paper – #US44413318, Sponsored by Seagate), November 2018, 6,

https://www.seagate.com/files/www-content/our-story/trends/files/idc-seagate-dataage
-whitepaper.pdf.

4. João Marques Lima, "Data centres of the world will consume 1/5 of Earth's power by 2025,"
   Data Economy, December 12, 2017, https://data-economy.com/data-centres-world-will-consume
   -1-5-earths-power-2025/.

5. Ryan Naraine, "Microsoft Makes Giant Anti-Spyware Acquisition," *eWEEK*, December 16,
   2004, http://www.eweek.com/news/microsoft-makes-giant-anti-spyware-acquisition.

6. The Microsoft antitrust saga illustrates many things, including the incredible length of time it
   can take for this type of scrutiny and enforcement to run its course if a company fails to address
   the concerns that garner the attention of government authorities. After resolving issues in the
   United States in the early 2000s, it took until December 2009 to reach the final major agreement in Brussels with the European Commission. European Commission, "Antitrust: Commission Accepts Microsoft Commitments to Give Users Browser Choice," December 16, 2009,
   http://europa.eu/rapid/press-release_IP-09-1941_en.htm.

   From start to finish, the many investigations into and lawsuits against Microsoft ran
   almost three decades. The company's antitrust issues began in June 1990, when the Federal
   Trade Commission opened what became a well-publicized review of marketing, licensing, and
   distribution practices for the Windows operating system. Andrew I. Gavil and Harry First, *The
   Microsoft Antitrust Cases: Competition Policy for the Twenty-First Century* (Cambridge, MA: The
   MIT Press, 2014.) The cases took many twists and turns, with the last lawsuits resolved more
   than twenty-eight years later, on December 21, 2018. Reflecting the broadening nature of
   what became in some ways the first truly global antitrust controversy, with investigations and
   proceedings in twenty-seven countries, the final cases were consumer class action matters in
   three Canadian provinces—Quebec, Ontario, and British Columbia.

   While three decades might at first blush seem shockingly long for a technology policy
   issue, in many ways it is more typical than most would imagine for major antitrust issues. In
   1999, as Microsoft was in the throes of its biggest case, I spent time studying the big antitrust
   battles of the twentieth century, including how the companies and their CEOs had addressed
   them. This included companies like Standard Oil, U.S. Steel, IBM, and AT&T, all companies
   that had defined the leading technologies of their time. The US government brought its first
   antitrust case against AT&T in 1913, and despite respites between major cases, the issues did
   not end until 1982, when the company agreed to be broken up to resolve the third major antitrust action against it. Similarly, IBM confronted the government in its first major lawsuit in
   1932, and disputes regarding its mainframe dominance continued until it settled a major case
   with the European Commission in 1984. It took another decade for IBM's mainframe dominance to subside to the point where it felt it could petition authorities in Washington, DC, and
   Brussels to end its settlement oversight. Tom Buerkle, "IBM Moves to Defend Mainframe
   Business in EU," *New York Times*, July 8, 1994, https://www.nytimes.com/1994/07/08/business
   /worldbusiness/IHT-ibm-moves-to-defend-mainframe-business-in-eu.html.

   The length of these battles provided a lesson that shaped my thinking about how technology companies need to approach antitrust and other regulatory issues. It led me to conclude
   at the time that successful tech companies needed to chart a proactive course to engage with
   the authorities, strengthen relationships, and, ultimately, construct more stable arrangements
   with governments.

CHAPTER 1: SURVEILLANCE

1. Glenn Greenwald, "NSA Collecting Phone Records of Millions of Verizon Customers Daily,"
   Guardian, June 6, 2013, https://www.theguardian.com/world/2013/jun/06/nsa-phone-records
   -verizon-court-order.

2. Glenn Greenwald and Ewen MacAskill, "NSA Prism Program Taps In to User Data of Apple,
   Google and Others," *Guardian*, June 7, 2013, https://www.theguardian.com/world/2013/jun/06
   /us-tech-giants-nsa-data.

3. Benjamin Dreyfuss and Emily Dreyfuss, "What Is the NSA's PRISM Program? (FAQ),"
   CNET, June 7, 2013, https://www.cnet.com/news/what-is-the-nsas-prism-program-faq/.

4. James Clapper, who was the national intelligence director at the time, would later describe the
   program as an "internal government computer system used to facilitate the government's
   statutorily authorized collection of foreign intelligence information from electronic

communication service providers under court supervision." Robert O'Harrow Jr., Ellen Nakashima, and Barton Gellman, "U.S., Company Officials: Internet Surveillance Does Not Indiscriminately Mine Data," *Washington Post*, June 8, 2013, https://www.washingtonpost.com /world/national-security/us-company-officials-internet-surveillance-does-not-indiscriminately -mine-data/2013/06/08/5b3bb234-d07d-11e2-9f1a-1a7cdee20287_story.html?utm_term= .b5761610edb1.

5. Glenn Greenwald, Ewen MacAskill, and Laura Poitras, "Edward Snowden: The Whistleblower Behind the NSA Surveillance Revelations," *Guardian*, June 11, 2013, https://www.theguard ian.com/world/2013/jun/09/edward-snowden-nsa-whistleblower-surveillance.

6. Michael B. Kelley, "NSA: Snowden Stole 1.7 Million Classified Documents and Still Has Access to Most of Them," *Business Insider*, December 13, 2013, https://www.businessinsider.com /how-many-docs-did-snowden-take-2013-12.

7. Ken Dilanian, Richard A. Serrano, and Michael A. Memoli, "Snowden Smuggled Out Data on Thumb Drive, Officials Say," *Los Angeles Times*, June 13, 2013, http://articles.latimes.com/2013 /jun/13/nation/la-na-nsa-leaks-20130614.

8. Nick Hopkins, "UK Gathering Secret Intelligence Via Covert NSA Operation," *Guardian*, June 7, 2013, https://www.theguardian.com/technology/2013/jun/07/uk-gathering-secret-intelligence -nsa-prism; see also Mirren Gidda, "Edward Snowden and the NSA Files—Timeline," *Guardian*, August 21, 2013, https://www.theguardian.com/world/2013/jun/23/edward-snowden-nsa -files-timeline.

9. William J. Cuddihy, *The Fourth Amendment: Origins and Meaning, 1602–1791* (Oxford: Oxford University Press, 2009), 441.

10. Ibid., 442.

11. Ibid., 459.

12. Frederick S. Lane, *American Privacy: The 400-Year History of Our Most Contested Right* (Boston: Beacon Press, 2009), 11.

13. David Fellman, *The Defendant's Rights Today* (Madison: University of Wisconsin Press, 1976), 258.

14. William Tudor, *The Life of James Otis, of Massachusetts: Containing Also, Notices of Some Contemporary Characters and Events, From the Year 1760 to 1775* (Boston: Wells and Lilly, 1823), 87–88. Adams recalled the impact of Otis's words on the people of Massachusetts the day after the nation's founders voted for independence in Philadelphia on July 2, 1776. Adams woke early to pen a letter to his wife, Abigail, recalling Otis's importance. Brad Smith, "Remembering the Third of July," *Microsoft on the Issues* (blog), Microsoft, July 3, 2014, https://blogs.microsoft .com/on-the-issues/2014/07/03/remembering-the-third-of-july/

15. David McCullough, *John Adams* (New York: Simon & Schuster, 2001), 62. William Cranch, *Memoir of the Life, Character, and Writings of John Adams* (Washington, DC: Columbian Institute, 1827), 15. Interestingly, Otis's advocacy and Adams's recognition of its importance have continued to influence American public policy and law into our current day. US Chief Justice John Roberts first quoted their words in 2014 when he wrote the Supreme Court's unanimous opinion requiring that law enforcement secure a search warrant before inspecting the contents of a suspect's smartphone. *Riley v. California*, 573 U.S. _ (2014), https://www.supremecourt .gov/opinions/13pdf/13-132_8l9c.pdf, at 27–28. Roberts did so again in 2018 when he wrote for a majority of the court that the police similarly needed a warrant to access cell phone location records. *Carpenter v. United States*, No. 16-402, 585 U.S. (2017), https://www.supremecourt.gov /opinions/17pdf/16-402_h315.pdf, at 5.

16. Thomas K, Clancy, *The Fourth Amendment: Its History and Interpretation* (Durham, NC: Carolina Academic Press, 2014), 69–74.

17. US Constitution, amendment IV.

18. Brent E. Turvey and Stan Crowder, *Ethical Justice: Applied Issues for Criminal Justice Students and Professionals* (Oxford: Academic Press, 2013), 182–83.

19. Ex parte Jackson, 96 U.S. 727 (1878).

20. Cliff Roberson, *Constitutional Law and Criminal Justice*, second edition (Boca Raton, FL: CRC Press, 2016), 50; Clancy, *The Fourth Amendment*, 91–104.

21. Charlie Savage, "Government Releases Once-Secret Report on Post-9/11 Surveillance," *New York Times*, April 24, 2015, https://www.nytimes.com/interactive/2015/04/25/us/25stellarwind-ig -report.html.

22. Terri Diane Halperin, *The Alien and Sedition Acts of 1798: Testing the Constitution* (Baltimore: John Hopkins University Press, 2016), 42–43.

23. Ibid., 59–60.
24. David Greenberg, "Lincoln's Crackdown," *Slate*, November 30, 2001, https://slate.com/news-and-politics/2001/11/lincoln-s-suspension-of-habeas-corpus.html.
25. T. A. Frail, "The Injustice of Japanese-American Internment Camps Resonates Strongly to This Day," *Smithsonian*, January 2017, https://www.smithsonianmag.com/history/injustice-japanese-americans-internment-camps-resonates-strongly-180961422/.
26. Barton Gellman and Ashkan Soltani, "NSA Infiltrates Links to Yahoo, Google Data Centers Worldwide, Snowden Documents Say," *Washington Post*, October 30, 2013, https://www.washingtonpost.com/world/national-security/nsa-infiltrates-links-to-yahoo-google-data-centers-worldwide-snowden-documents-say/2013/10/30/e51d661e-4166-11e3-8b74-d89d714ca4dd_story.html?noredirect=on&utm_term=.5c2f99fcc376.
27. "Evidence of Microsoft's Vulnerability," *Washington Post*, November 26, 2013, https://www.washingtonpost.com/apps/g/page/world/evidence-of-microsofts-vulnerability/621/.
28. Craig Timberg, Barton Gellman, and Ashkan Soltani, "Microsoft, Suspecting NSA Spying, to Ramp Up Efforts to Encrypt Its Internet Traffic," *Washington Post*, November 26, 2013, https://www.washingtonpost.com/business/technology/microsoft-suspecting-nsa-spying-to-ramp-up-efforts-to-encrypt-its-internet-traffic/2013/11/26/44236b48-56a9-11e3-8304-caf30787c0a9_story.html?utm_term=.69201c4e9ed8.
29. "Roosevelt Room," White House Museum, accessed February 20, 2019, http://www.whitehousemuseum.org/west-wing/roosevelt-room.htm.
30. A couple of press reports focused on Pincus's suggestion that Obama pardon Snowden. Seth Rosenblatt, "'Pardon Snowden,' One Tech Exec Tells Obama, Report Says," Cnet, December 18, 2013, https://www.cnet.com/news/pardon-snowden-one-tech-exec-tells-obama-report-says/; Dean Takahashi, "Zynga's Mark Pincus Asked Obama to Pardon NSA Leaker Edward Snowden," *VentureBeat*, December 19, 2013, https://venturebeat.com/2013/12/19/zyngas-mark-pincus-asked-president-obama-to-pardon-nsa-leaker-edward-snowden/.
31. "Transcript of President Obama's Jan. 17 Speech on NSA Reform," *Washington Post*, January 17, 2014, https://www.washingtonpost.com/politics/full-text-of-president-obamas-jan-17-speech-on-nsa-reforms/2014/01/17/fa33590a-7f8c-11e3-9556-4a4bf7bcbd84_story.html?utm_term=.c8d2871c4f72.

CHAPTER 2: TECHNOLOGY AND PUBLIC SAFETY

1. "Reporter Daniel Pearl Is Dead, Killed by His Captors in Pakistan," *Wall Street Journal*, February 24, 2002, http://online.wsj.com/public/resources/documents/pearl-022102.htm.
2. Electronic Communications Privacy Act of 1986, Public Law 99-508, 99th Cong., 2d sess. (October 21, 1986), 18 U.S.C. § 2702.b.
3. Electronic Communications Privacy Act of 1986, Public Law 99-508, 99th Cong., 2d sess. (October 21, 1986), 18 U.S.C. Chapter 121 §§ 2701 et seq.
4. Electronic Communications Privacy Act of 1986, Public Law 99-508, 99th Cong., 2d sess. (October 21, 1986), 18 U.S.C. § 2705.b.
5. "Law Enforcement Requests Report," Corporate Social Responsibility, Microsoft, last modified June 2018, https://www.microsoft.com/en-us/about/corporate-responsibility/lerr/.
6. "Charlie Hebdo Attack: Three Days of Terror," *BBC News*, January 14, 2015, https://www.bbc.com/news/world-europe-30708237.
7. "Al-Qaeda in Yemen Claims Charlie Hebdo Attack," *Al Jezeera*, 14 Jan 2015, https://www.aljazeera.com/news/middleeast/2015/01/al-qaeda-yemen-charlie-hebdo-paris-attacks-201511410323361511.html.
8. Ibid.
9. "Paris Attacks: Millions Rally for Unity in France," *BBC News*, January 11, 2015, https://www.bbc.com/news/world-europe-30765824.
10. Alissa J. Rubin, "Paris One Year On," *New York Times*, November 12, 2016, https://www.nytimes.com/2016/11/13/world/europe/paris-one-year-on.html.
11. "Brad Smith: New America Foundation: 'Windows Principles,'" *Stories* (blog), Microsoft, July 19, 2006, https://news.microsoft.com/speeches/brad-smith-new-america-foundation-windows-principles/.
12. It took several months to develop a clear set of principles. The effort was led by Horacio Gutierrez, then the most senior product lawyer at Microsoft and now the general counsel with wide-ranging business responsibilities at Spotify. He partnered with Mark Penn, a former

Clinton official with a keen marketing sense. Horacio assembled an internal team that spanned the various parts of the company, and he enlisted a team from the Boston Consulting Group to help us survey customers to learn what they valued most. Horacio and the team developed the four principles, which I unveiled publicly as our cloud commitments in July 2015. Brad Smith, "Building a Trusted Cloud in an Uncertain World," Microsoft Worldwide Partner Conference, Orlando, July 15, 2015, video of keynote, https://www.youtube.com/watch?v=RkAwAj1Z9rg.

13. "Responding to Government Legal Demands for Customer Data," *Microsoft on the Issues* (blog), Microsoft, July 16, 2013, https://blogs.microsoft.com/on-the-issues/2013/07/16/responding-to -government-legal-demands-for-customer-data/.

14. *United States v. Jones*, 565 U.S. 400 (2012), https://www.law.cornell.edu/supremecourt/text /10-1259.

15. Ibid., 4.

16. *Riley v. California*, 573 U.S. _ (2014).

17. Ibid., 20.

18. Ibid., 21.

19. Steve Lohr, "Microsoft Sues Justice Department to Protest Electronic Gag Order Statute," *New York Times*, April 14, 2016, https://www.nytimes.com/2016/04/15/technology/microsoft -sues-us-over-orders-barring-it-from-revealing-surveillance.html?_r=0.

20. Brad Smith, "Keeping Secrecy the Exception, Not the Rule: An Issue for Both Consumers and Businesses," *Microsoft on the Issues* (blog), Microsoft, April 14, 2016, https://blogs.microsoft .com/on-the-issues/2016/04/14/keeping-secrecy-exception-not-rule-issue-consumers -businesses/.

21. Rachel Lerman, "Long List of Groups Backs Microsoft in Case Involving Digital-Data Privacy," *Seattle Times*, September 2, 2016, https://www.seattletimes.com/business/microsoft /ex-federal-law-officials-back-microsoft-in-case-involving-digital-data-privacy/?utm_source =RSS&utm_medium=Referral&utm_campaign=RSS_all.

22. Cyrus Farivar, "Judge Sides with Microsoft, Allows 'Gag Order' Challenge to Advance," *Ars Technica*, February 9, 2017, https://arstechnica.com/tech-policy/2017/02/judge-sides-with-micro soft-allows-gag-order-challenge-to-advance/.

23. Brad Smith, "DOJ Acts to Curb the Overuse of Secrecy Orders. Now It's Congress' Turn," *Microsoft on the Issues* (blog), Microsoft, October 23, 2016, https://blogs.microsoft.com/on-the -issues/2017/10/23/doj-acts-curb-overuse-secrecy-orders-now-congress-turn/.

CHAPTER 3: PRIVACY

1. Tony Judt, *Postwar: A History of Europe since 1945* (New York: Penguin, 2006), 697.

2. Anna Funder, *Stasiland: True Stories from Behind the Berlin Wall* (London: Granta, 2003), 57.

3. Brad Smith and Carol Ann Browne, "Lessons on Protecting Privacy," *Today in Technology* (video blog), Microsoft, accessed April 7, 2019, https://blogs.microsoft.com/today-in-tech/vid eos/.

4. Jake Brutlag, "Speed Matters," Google AI Blog, June 23 2009, https://ai.googleblog.com/2009 /06/speed-matters.html.

5. The tension reached one peak in 1807, when the British HMS *Leopard*, sailing off the Virginia capes, demanded that the USS *Chesapeake* turn over four members of the *Chesapeake*'s crew, believed to be British deserters. When the *Chesapeake* refused, the *Leopard* fired seven broad-sides and forced the American ship to strike its flag. The *Leopard* recovered the four crewmen and the *Chesapeake* limped back to port. Jefferson closed American ports to British warships and declared a trade embargo. Craig L. Symonds, *The U.S. Navy: A Concise History* (Oxford: Oxford University Press, 2016), 21.

   Not surprisingly, the halt to trade hurt the United States as well as Great Britain. As one historian remarked, "Jefferson's embargo struck so hard throughout the nation that many countrymen concluded that he had declared war on them, not the British." A.J. Langguth, *Union 1812: The Americans Who Fought the Second War of Independence* (New York: Simon & Schuster, 2006), 134. Congress repealed the embargo three days before James Madison as-sumed the Presidency in 1809, but it continued to restrict trade with Great Britain. The British continued to use press gangs, and in 1811 a British frigate stopped and removed an American sailor from a merchant ship in sight of the New Jersey shore. Symonds, 23.

6. "Treaties, Agreements, and Asset Sharing," U.S. Department of State, https://www.state.gov /j/inl/rls/nrcrpt/2014/vol2/222469.htm.

7. Drew Mitnick, "The urgent need for MLAT reform," *Access Now*, September 12, 2014, https://www.accessnow.org/the-urgent-needs-for-mlat-reform/.

8. By coincidence, another judicial clerk arrived at the same time with a personal computer. His name was Eben Moglen, and he worked for a judge across the corridor on the twenty-second floor in Foley Square. We often chatted about our common interest in PCs. Eben would go on to become an impressive academic and leader of the open-source movement, becoming a professor of law at Columbia University and the chairman of the Software Freedom Law Center. At times in the early 2000s we found ourselves on opposing sides of legal debates involving software intellectual property issues.

9. The legislative process kicked off in earnest in 2015 when a bipartisan group of three Senators and two Representatives introduced the LEADS Act, short for Law Enforcement Access to Data Stored Abroad. It was co-sponsored in the Senate by Orrin Hatch, Chris Coons, and Dean Heller, and in the House by Tom Marino and Suzan DelBene. Patrick Maines, "The LEADS Act and Cloud Computing," *The Hill*, March 30, 2015, https://thehill.com/blogs/pundits-blog/technology/237328-the-leads-act-and-cloud-computing.

10. There was naturally a long and winding road between our initial loss before Judge Francis in 2014 and our arrival at the steps of the Supreme Court in 2018. We lost the next round of litigation at the District Court level before Chief Judge Loretta Preska, who ruled against us in July 2014. It was a lively two-hour argument, with the government's lawyer relying on the fact that the US government could force companies to produce their business records from around the world. Our team made what we always regarded as one of our fundamental points, that other people's emails didn't belong to us and were not our business records to treat however we wished. But Judge Preska didn't buy it and surprised us by giving her ruling orally in the courtroom as the oral argument ended. Ellen Nakashima, "Judge Orders Microsoft to Turn Over Data Held Overseas," *Washington Post*, July 31, 2014, https://www.washingtonpost.com/world/national-security/judge-orders-microsoft-to-turn-over-data-held-overseas/2014/07/31/b07c4952-18d4-11e4-9e3b-7f2f110c6265_story.html?utm_term=.e913e692474e. As the *Post* put it, "The judge's ruling probably will prompt more expressions of outrage from foreign officials, especially in the European Union, about the potential for intrusion into their sovereignty." That in fact was the case.

　　The next round took us to the Court of Appeals for the Second Circuit, which considers all appeals from district court decisions in New York, Connecticut, and Vermont. As we prepared for this step and in part with an eye on the ultimate need for legislation, we decided to try to broaden the public discussion and enlist more voices in it. We embarked on a major recruiting drive to ask groups to support us by filing amicus—so-called "friend of the court"—briefs. We quickly garnered support from a wide variety of organizations but worried about trying to break through in a crowded news cycle.

　　We came up with an idea: Why not produce our own broadcast show to bring the issues and support to life? We could create short videos to show a data center and explain the issues in more approachable terms. We could invite experts to break down the issues, explain why people needed to pay attention and push for reform. The event could take place at Microsoft's new offices in New York. The press could attend in person while we live-streamed the session on the web, beamed out with an additional important audience in mind: the US Congress.

　　We concluded that we needed a respected journalist to learn about the issues and serve as moderator. I had gotten to know Charlie Gibson, the famed and respected former ABC news anchor, as a member of the Princeton University Board of Trustees. Happily, he agreed to serve in the role as long as he could ask people hard questions as serious journalists would expect. We readily agreed.

　　On a frigid morning in December 2014, we broadcast our electronic privacy program from Microsoft's New York office in Times Square. We announced the filing of amicus briefs from groups that included twenty-eight tech and media companies, twenty-three trade associations and advocacy groups, and thirty-five leading computer scientists. And to top it off, there was a supportive brief from the Government of Ireland itself. As I announced the filings, I joked that this was the first time anyone could remember that the ACLU and Fox News had worked together and were on the same side. Video of the event: https://ll.ms-studiosmedia.com/events/2014/1412/ElectronicPrivacy/live/ElectronicPrivacy.html. The event accomplished what we wanted, generating news coverage across the country and around the world. And perhaps most importantly, with the strange bedfellows that had come together to endorse our approach, more people in Congress started to take notice.

In July 2016, more than seven months after the oral argument in New York, a unanimous three-judge panel in the Second Circuit ruled in our favor. Brad Smith, "Our Search Warrant Case: An Important Decision for People Everywhere," *Microsoft on the Issues* (blog), Microsoft, July 14, 2016, https://blogs.microsoft.com/on-the-issues/2016/07/14/search-warrant-case-import ant-decision-people-everywhere/. The Department of Justice then successfully persuaded the Supreme Court to consider the case, which brought us to the steps of that courthouse in 2018.

11. *Microsoft Corp. v. AT&T Corp.*, 550 U.S. 437 (2007).
12. Official Transcript, *Microsoft Corp. v. AT&T Corp.*, February 21, 2007.
13. Clarifying Lawful Overseas Use of Data Act of 2018, H.R. 4943, 115th Cong. (2018).
14. Brad Smith, "The CLOUD Act Is an Important Step Forward, but Now More Steps Need to Follow," *Microsoft on the Issues* (blog), Microsoft, April 3, 2018, https://blogs.microsoft.com/on -the-issues/2018/04/03/the-cloud-act-is-an-important-step-forward-but-now-more-steps -need-to-follow/.
15. Derek B. Johnson, "The CLOUD Act, One Year On," *FCW: The Business of Federal Technology*, April 8, 2019, https://fcw.com/articles/2019/04/08/cloud-act-turns-one.aspx.

CHAPTER 4: CYBERSECURITY

1. "St Bartholomew's Hospital during World War Two," BBC, December 19, 2005, https://www .bbc.co.uk/history/ww2peopleswar/stories/10/a7884110.shtml.
2. "What Does NHS England Do?" NHS England, accessed November 14, 2018, https://www .england.nhs.uk/about/about-nhs-england/.
3. Kim Zetter, "Sony Got Hacked Hard: What We Know and Don't Know So Far," *Wired*, December 3, 2014, https://www.wired.com/2014/12/sony-hack-what-we-know/.
4. Bill Chappell, "WannaCry Ransomware: What We Know Monday," NPR, May 15, 2017, https://www.npr.org/sections/thetwo-way/2017/05/15/528451534/wannacry-ransomware -what-we-know-monday.
5. Nicole Perlroth and David E. Sanger, "Hackers Hit Dozens of Countries Exploiting Stolen N.S.A. Tool," *New York Times*, May 12, 2017, https://www.nytimes.com/2017/05/12/world/eu rope/uk-national-health-service-cyberattack.html.
6. Bruce Schneier, "Who Are the Shadow Brokers?" *The Atlantic*, 23 May 2017. https://www.the atlantic.com/technology/archive/2017/05/shadow-brokers/527778/.
7. Nicole Perlroth and David E. Sanger, "Hackers Hit Dozens of Countries Exploiting Stolen N.S.A. Tool," *New York Times*, May 12, 2017, https://www.nytimes.com/2017/05/12/world/eu rope/uk-national-health-service-cyberattack.html.
8. Brad Smith, "The Need for Urgent Collective Action to Keep People Safe Online: Lessons from Last Week's Cyberattack," *Microsoft on the Issues* (blog), Microsoft, May 14 2017, https:// blogs.microsoft.com/on-the-issues/2017/05/14/need-urgent-collective-action-keep-people -safe-online-lessons-last-weeks-cyberattack/.
9. Choe Sang-Hun, David E. Sanger, and William J. Broad, "North Korean Missile Launch Fails, and a Show of Strength Fizzles," *New York Times*, April 15, 2017, https://www.nytimes.com /2017/04/15/world/asia/north-korea-missiles-pyongyang-kim-jong-un.html.
10. Lily Hay Newman, "How an Accidental 'Kill Switch' Slowed Friday's Massive Ransomware Attack," *Wired*, May 13, 2017, https://www.wired.com/2017/05/accidental-kill-switch-slowed -fridays-massive-ransomware-attack/.
11. Andy Greenberg, "The Untold Story of NotPetya, the Most Devastating Cyberattack in History," *Wired*, August 22, 2018, https://www.wired.com/story/notpetya-cyberattack-ukraine -russia-code-crashed-the-world/.
12. Ibid.; Stilgherrian, "Blaming Russia for NotPetya Was Coordinated Diplomatic Action," *ZDNet*, April 12, 2018, https://www.zdnet.com/article/blaming-russia-for-notpetya-was-coordinated -diplomatic-action.
13. Josh Fruhlinger, "Petya Ransomware and NotPetya Malware: What You Need to Know Now," October 17, 2017, https://www.csoonline.com/article/3233210/petya-ransomware-and-notpetya -malware-what-you-need-to-know-now.html.
14. Greenberg, "The Untold Story of NotPetya."
15. Microsoft, "RSA 2018: The Effects of NotPetya," YouTube video, 1:03, produced by Brad Smith, Carol Ann Browne, and Thanh Tan, April 17, 2018, https://www.youtube.com/watch ?time_continue=1&v=QVhqNNO0DNM.

16. Andy Sharp, David Tweed, and Toluse Olorunnipa, "U.S. Says North Korea Was Behind Wan-naCry Cyberattack," *Bloomberg*, December 18, 2017, https://www.bloomberg.com/news/arti cles/2017-12-19/u-s-blames-north-korea-for-cowardly-wannacry-cyberattack.

CHAPTER 5: PROTECTING DEMOCRACY

1. Max Farrand, ed., *The Records of the Federal Convention of 1787* (New Haven, CT: Yale University Press, 1911), 3:85.
2. The Digital Crimes Unit has evolved almost constantly since we first moved beyond the anti-counterfeiting work that initially led us to recruit investigators and former prosecutors to address criminal activity involving new technology. One key turning point came in the early 2000s, when the chief of the Toronto police force came to Redmond. He was on a mission to persuade us to make a major investment to help police fight child pornography and exploitation around the world. As I walked downstairs to meet with him in a conference room, I was convinced that we had no room in our budget to take on this new mission. I left the meeting ninety minutes later convinced that we had no choice but to help put a dent in the online exploitation that remains one of the most horrific creations of the internet age. We cut spending elsewhere, and a new DCU team emerged that ever since has used a combination of technology and legal tactics to help protect children.

    Another moment came in 2008 when some of us visited Seoul, and the South Korean Government took me on a tour of its national cybercrime headquarters. We were impressed by their team and even more impressed by their cutting-edge facility, which was better than anything at our headquarters. We returned home and decided to create for the DCU a dedicated Cybercrime Center with the world's premier tools and resources for its work on our Redmond Campus. This includes separate, dedicated office space that can be used by visiting investigators and lawyers when the DCU pursues joint operations with law enforcement or other groups.

    By 2012, the DCU innovated in new ways to address cybercriminals' use of "botnets" to infect and take control of PCs around the world. Nick Wingfield and Nicole Perlroth, "Microsoft Raids Tackle Internet Crime," *New York Times*, March 26, 2012, https://www.nytimes .com/2012/03/26/technology/microsoft-raids-tackle-online-crime.html. DCU lawyer Richard Boscovich first developed the legal technique to take control of these groups' command and control servers based on arguments grounded in trademark infringement and the even older legal concept that protects against "trespass to chattels." I've always found it a bit amusing that we're protecting computers based on a legal doctrine first developed in England in part to protect cattle.

    More recently, the DCU has taken on the challenge of fighting the fraudulent and annoying telephone calls and other technology scams that seek to persuade people at home that their PC or smartphone is infected and they need to spend money to install new security software to fix it. Microsoft Assistant General Counsel Courtney Gregoire has led innovative work that has taken us to India and elsewhere around the world to address these problems at their source. Courtney Gregoire, "New Breakthroughs in Combatting Tech Support Scams," *Microsoft on the Issues* (blog), Microsoft, November 29, 2018, https://blogs.microsoft.com/on -the-issues/2018/11/29/new-breakthroughs-in-combatting-tech-support-scams/.
3. Brandi Buchman, "Microsoft Turns to Court to Break Hacker Ring," *Courthouse News Service*, August 10, 2016, https://www.courthousenews.com/microsoft-turns-to-court-to-break-hacker -ring/.
4. April Glaser, "Here Is What We Know About Russia and the DNC Hack," *Wired*, July 27, 2016, https://www.wired.com/2016/07/heres-know-russia-dnc-hack/.
5. Alex Hern, "Macron Hackers Linked to Russian-Affiliated Group Behind US Attack," *Guardian*, May 8, 2017, https://www.theguardian.com/world/2017/may/08/macron-hackers-linked-to -russian-affiliated-group-behind-us-attack.
6. Kevin Poulsen and Andrew Desiderio, "Russian Hackers' New Target: A Vulnerable Democratic Senator," *Daily Beast*, July 26, 2018, https://www.thedailybeast.com/russian-hackers -new-target-a-vulnerable-democratic-senator?ref=scroll.
7. Griffin Connolly, "Claire McCaskill Hackers Left Behind Clumsy Evidence That They Were Russian," *Roll Call*, August 23, 2018, https://www.rollcall.com/news/politics/mccaskill -hackers-evidence-russian.

8. Tom Burt, "Protecting Democracy with Microsoft AccountGuard," *Microsoft on the Issues* (blog), Microsoft, August 20, 2018, https://blogs.microsoft.com/on-the-issues/2018/08/20/pro tecting-democracy-with-microsoft-accountguard/.

9. Brad Smith, "We Are Taking New Steps Against Broadening Threats to Democracy," *Microsoft on the Issues* (blog), Microsoft, August 20, 2018, https://blogs.microsoft.com/on-the-issues /2018/08/20/we-are-taking-new-steps-against-broadening-threats-to-democracy/.

10. Brad Smith, "Microsoft Sounds Alarm on Russian Hacking Attempts," interview by Amna Nawaz, *PBS News Hour*, August 22, 2018, https://www.pbs.org/newshour/show/microsoft -sounds-alarm-on-russian-hacking-attempts.

11. "Moscow: Microsoft's Claim of Russian Meddling Designed to Exert Political Effect," *Sputnik International*, August 21, 2018, https://sputniknews.com/us/201808211067354346-us-microsoft -hackers/.

12. Tom Burt, "Protecting Democratic Elections Through Secure, Verifiable Voting," *Microsoft on the Issues* (blog), May 6, 2019, https://blogs.microsoft.com/on-the-issues/2019/05/06/protecting -democratic-elections-through-secure-verifiable-voting/.

CHAPTER 6: SOCIAL MEDIA

1. *Freedom Without Borders*, Permanent Exhibition, Vabamu Museum of Occupations and Freedom, Tallin, Estonia, https://vabamu.ee/plan-your-visit/permanent-exhibitions/freedom -without-borders.

2. Shortly after her birth, Olga's father, eager to escape the tumult and famine of Ukraine, accepted an appointment as chief surgeon at a hospital near the Moscow railway, with hopes of eventually immigrating north to Estonia. They never made it. The family's plans were cut short when Olga's mother, weakened by malnutrition, died suddenly from meningitis. Soon, her widowed father, who had dodged the Bolsheviks for several years, was imprisoned in a Siberian camp. At the age of two, Olga and her older brother, then seven, were fending for themselves, surviving on fish caught with makeshift nets in a nearby pond. Word of the abandoned children made its way to Tallinn where an uncle used his railway connections and aid from the Red Cross to bring the siblings safely to Estonia. Olga was taken into the arms of a charitable foster family, who raised her through two occupations and a global conflict, eventually sending her to the University of Tartu where she earned a medical degree. Toward the end of the Second World War, Olga fled with retreating German soldiers and made her way toward the American sector in Germany. Again, thanks to the kindness of strangers, Olga was lifted to safety—this time through the window of an overcrowded train just as it pulled from a station toward the city of Erlangen and into freedom. Ede Schank Tamkivi, "The Story of a Museum," Vabamu, Kistler-Ritso Eesti Sihtasutus, December 2018, 42.

3. Ede Schank Tamkivi, "The Story of a Museum," Vabamu, Kistler-Ritso Eesti Sihtasutus, December 2018, 42.

4. Damien McGuinness, "How a Cyber Attack Transformed Estonia," *BBC News*, April 27, 2017, https://www.bbc.com/news/39655415.

5. Rudi Volti, *Cars and Culture: The Life Story of a Technology* (Westport, CT: Greenwood Press, 2004), 40.

6. Ibid., 39.

7. Ibid.

8. Sherry Turkle, *Alone Together: Why We Expect More from Technology and Less from Each Other* (New York: Basic Books, 2011), 17.

9. Philip N. Howard, Bharath Ganesh, Dimitra Liotsiou, John Kelly, and Camille François, "The IRA, Social Media and Political Polarization in the United States, 2012–2018" (working paper, Computational Propaganda Research Project, University of Oxford, 2018), https://fas.org/irp /congress/2018_rpt/ira.pdf.

10. Ibid.

11. Ryan Lucas, "How Russia Used Facebook to Organize 2 Sets of Protesters," NPR, November 1, 2017, https://www.npr.org/2017/11/01/561427876/how-russia-used-facebook-to-organize-two -sets-of-protesters.

12. Deepa Seetharaman, "Zuckerberg Defends Facebook Against Charges It Harmed Political Discourse," *Wall Street Journal*, November 10, 2016, https://www.wsj.com/articles/zuckerberg -defends-facebook-against-charges-it-harmed-political-discourse-1478833876.

13. Chloe Watson, "The Key Moments from Mark Zuckerberg's Testimony to Congress," *Guardian*, April 11, 2018, https://www.theguardian.com/technology/2018/apr/11/mark-zuckerbergs-testimony-to-congress-the-key-moments.

14. Mark R. Warner, "Potential Policy Proposals for Regulation of Social Media and Technology Firms" (draft white paper, Senate Intelligence Committee, 2018), https://www.scribd.com/document/385137394/MRW-Social-Media-Regulation-Proposals-Developed.

15. When Congress passed the Communications Decency Act in 1996, it included section 230(c)(1), which states that "No provider or user of an interactive computer service shall be treated as the publisher or speaker of any information provided by another information content provider." 47 U.S.C. § 230, at https://www.law.cornell.edu/uscode/text/47/230. As one author has noted, "When first enacted by Congress, section 230 was intended to foster openness and innovation in the World Wide Web by giving websites broad legal protections and allowing the Internet to grow as a true marketplace of ideas. Advocates of online free speech at the time had argued that if controls were as tight on internet communication as with offline communication, the constant threat of litigation would intimidate individuals from weighing in on important issues of public concern." Marie K. Shanahan, *Journalism, Online Comments, and the Future of Public Discourse* (New York: Routledge, 2018), 90.

16. Ibid., 8.

17. Kevin Roose, "A Mass Murder of, and for, the Internet," *New York Times*, March 15, 2019, https://www.nytimes.com/2019/03/15/technology/facebook-youtube-christchurch-shooting.html.

18. Ibid.

19. Matt Novak, "New Zealand's Prime Minister Says Social Media Can't Be 'All Profit, No Responsibility,'" *Gizmodo*, March 19, 2019, https://gizmodo.com/new-zealands-prime-minister-says-social-media-cant-be-a-1833398451.

20. Ibid.

21. Milestones: Westinghouse Radio Station KDKA, 1920, *Engineering and Technology History Wiki*, https://ethw.org/Milestones:Westinghouse_Radio_Station_KDKA,_1920.

22. Stephen Smith, "Radio: The Internet of the 1930s," *American RadioWorks*, November 10, 2014, http://www.americanradioworks.org/segments/radio-the-internet-of-the-1930s/.

23. Ibid.

24. Vaughan Bell, "Don't Touch That Dial! A History of Media Technology Scares, from the Printing Press to Facebook," *Slate*, February 15, 2010, https://slate.com/technology/2010/02/a-history-of-media-technology-scares-from-the-printing-press-to-facebook.html.

25. Vincent Pickard, "The Revolt Against Radio: Postwar Media Criticism and the Struggle for Broadcast Reform," in *Moment of Danger: Critical Studies in the History of U.S. Communication Since World War II* (Milwaukee: Marquette University Press, 2011), 35–56.

26. Ibid., 36.

27. Vincent Pickard, "The Battle Over the FCC Blue Book: Determining the Role of Broadcast Media in a Democratic Society, 1945–1948," *Media, Culture & Society* 33(2), 171–91, https://doi.org/10.1177/0163443710385504. As recognized by another scholar, "The Blue Book was not simply a remarkable regulatory moment in the history of the FCC; it was also the catalyst for the most widespread public discussion of advertising and broadcasting in American history." Michael Socolow, "Questioning Advertising's Influence over American Radio: The Blue Book Controversy of 1945–1947," *Journal of Radio Studies* 9(2), 282, 287.

28. As Socolow has observed, "The Blue Book led to a new consciousness of responsibility within the industry." Ibid., 297. Among specific developments that followed, CBS and NBC adopted stringent self-regulatory codes. CBS established a documentary unit, which led NBC to launch a new series to compete with it. Ibid., 297–98.

29. The Parliament of the Commonwealth of Australia, "Criminal Code Amendment (Sharing of Abhorrent Violent Material) Bill 2019, A Bill for an Act to Amend the Criminal Code Act 1995, and for Related Purposes," https://parlinfo.aph.gov.au/parlInfo/download/legislation/bills/s1201_first-senate/toc_pdf/1908121.pdf;fileType=application%2F.pdf; Jonathan Shieber, "Australia Passes Law to Hold Social Media Companies Responsible for 'Abhorrent Violent Material,'" *TechCrunch*, April 4, 2019, https://techcrunch.com/2019/04/04/australia-passes-law-to-hold-social-media-companies-responsible-for-abhorrent-violent-material/. I spent a day in Canberra following my two days in Wellington, just eight days before the Australian law was passed. Reflecting the speed involved, at that point a piece of legislation hadn't yet even been unveiled.

30. When I was in Canberra the week before the new law was passed, I sought to make the case for strong but more deliberative action. As I said to the *Australian Financial Review*, "I think that governments do need to start moving faster on technology issues, but one always needs to be very careful not to move faster than the speed of thought." I quickly added, "Not surprisingly, I'm not going to be the foremost advocate of sending myself or my colleagues in other companies to prison. I think that might have a chilling effect on international travel, that actually helps us understand what the world's people need from our products." Paul Smith, "Microsoft President Says Big Tech Regulation Must Learn from History," *The Australian Financial Review*, April 2, 2019, https://www.afr.com/technology/technology-companies/microsoft-presi dent-says-big-tech-regulation-must-learn-from-history-20190329-p518v2.

31. Warner, 9.

32. HM Government, *Online Harms White Paper*, April 2019, 7, https://assets.publishing.service .gov.uk/government/uploads/system/uploads/attachment_data/file/793360/Online_Harms _White_Paper.pdf.

33. "Restoring Trust & Accountability," *NewsGuard*, last modified 2019, https://www.newsguard tech.com/how-it-works/.

34. Ibid.

35. George C. Herring, *From Colony to Superpower: U.S. Foreign Relations Since 1776* (Oxford: Oxford University Press, 2008), 72.

36. Ironically, the Jacobins who rose to power during the French Revolution soon revoked Genêt's papers and called for his arrest and execution. "In a stunning show of magnanimity, Washington granted Genêt asylum, and the Frenchman who had been willing to overthrow the first government of the United States declared his allegiance to the American flag, renounced his French citizenship, married the daughter of New York governor George Clinton, and retired to a farm in Jamaica, Long Island. He died a high-living hypocrite who came to love the land he had sought to undermine as an arrogant youth. In another country he would have been hanged." John Avalon, *Washington's Farewell: The Founding Father's Warning to Future Generations* (New York: Simon & Schuster, 2017), 66.

37. George Washington, "Washington's Farewell Address of 1796," Avalon Project, Lillian Goldman Law Library, Yale Law School, http://avalon.law.yale.edu/18th_century/washing.asp.

CHAPTER 7: DIGITAL DIPLOMACY

1. Robbie Gramer, "Denmark Creates the World's First Ever Digital Ambassador," *Foreign Policy*, January 27, 2017, https://foreignpolicy.com/2017/01/27/denmark-creates-the-worlds -first-ever-digital-ambassador-technology-europe-diplomacy/.

2. Henry V. Poor, *Manual of the Railroads of the United States for 1883* (New York: H. V. & H. W. Poor, 1883), iv.

3. James W. Ely Jr., *Railroads & American Law* (Lawrence: University Press of Kansas, 2003). Another particularly good book that charts the long arc of technology regulation for railroads is Steven W. Usselman, *Regulating Railroad Innovation* (Cambridge, UK: Cambridge University Press, 2002).

4. Brad Smith, "Trust in the Cloud in Tumultuous Times," March 1, 2016, RSA Conference, Moscone Center San Francisco, video, 30:35, https://www.rsaconference.com/events/us16 /agenda/sessions/2750/trust-in-the-cloud-in-tumultuous-times.

5. Siemens AG, *Charter of Trust on Cybersecurity*, July 2018, https://www.siemens.com/content /dam/webassetpool/mam/tag-siemens-com/smdb/corporate-core/topic-areas/digitalization/cy bersecurity/charteroftrust-standard-presentation-july2018-en-1.pdf.

6. Brad Smith, "The Need for a Digital Geneva Convention," *Microsoft on the Issues* (blog), Microsoft, February 14, 2017, https://blogs.microsoft.com/on-the-issues/2017/02/14/need -digital-geneva-convention/.

7. Elizabeth Weise, "Microsoft Calls for 'Digital Geneva Convention," *USA Today*, February 14, 2017, https://www.usatoday.com/story/tech/news/2017/02/14/microsoft-brad-smith-digital-geneva -convention/97883896/.

8. Brad Smith, "We Need to Modernize International Agreements to Create a Safer Digital World," *Microsoft on the Issues* (blog), Microsoft, November 10, 2017, https://blogs.microsoft .com/on-the-issues/2017/11/10/need-modernize-international-agreements-create-safer-digital -world/.

9. A good firsthand account was authored in 1989 by Paul Nitze, one of the principal arms nego-tiators of the Cold War era. Paul Nitze, *From Hiroshima to Glasnost: At the Center of Decision, A Memoir* (New York: Grove Weidenfeld, 1989).

10. David Smith, "Movie Night with the Reagans: WarGames, Red Dawn . . . and Ferris Bueller's Day Off," *Guardian*, March 3, 2018, https://www.theguardian.com/us-news/2018/mar/03/movie-night-with-the-reagans.

11. *WarGames*, directed by John Badham (Beverly Hills: United Artists, 1983).

12. Fred Kaplan, *Dark Territory: The Secret History of Cyber War* (New York: Simon & Schuster, 2016), 1–2.

13. Seth Rosenblatt, "Where Did the CFAA Come From, and Where Is It Going?" *The Parallax*, March 16, 2016, https://the-parallax.com/2016/03/16/where-did-the-cfaa-come-from-and-where-is-it-going/.

14. Michael McFaul, *From Cold War to Hot Peace: An American Ambassador in Putin's Russia* (Boston: Houghton Mifflin Harcourt, 2018).

15. Paul Scharre, *Army of None: Autonomous Weapons and the Future of War* (New York: W. W. Norton, 2018), 251.

16. The International Committee of the Red Cross, or ICRC, plays a vital role today in every aspect of the implementation and promotion of compliance with the Geneva Conventions. This is despite the fact that, as two legal scholars have recognized, "a wide gap separates the meagre language of the [Geneva] Conventions' provisions mandating ICRC operations from the broad perception and exercise in practice of that mandate by the ICRC." Rotem Giladi and Steven Ratner, "The Role of the International Committee of the Red Cross," in Andrew Clapham, Paola Gaeta, and Marco Sassoli, eds., *The 1949 Geneva Conventions: A Commentary* (Oxford: Oxford University Press, 2015). The ICRC's success speaks to the uniquely credible role that a nongovernmental organization can play if it can successfully establish its credibility in a sustained way over an extended period of time.

17. Jeffrey W. Knopf, "NGOs, Social Movements, and Arms Control," in *Arms Control: History, Theory, and Policy, Volume 1: Foundations of Arms Control*, ed. Robert E. Williams Jr. and Paul R. Votti (Santa Barbara: Praeger, 2012), 174–75.

18. Bruce D. Berkowitz, *Calculated Risks: A Century of Arms Control, Why It Has Failed, and How It Can Be Made to Work* (New York: Simon and Schuster, 1987), 156.

19. Arguably the most influential such effort has involved an international group of experts that twice has come together at the NATO Cooperative Cyber Defence Centre of Excellence in Tallinn, Estonia. The group's most recent work resulted in an influential work with a less than dramatic title, the Tallinn Manual 2.0. It contains 154 rules that the experts concluded represent "the international law governing cyber warfare." Michael N. Schmitt, ed., *Tallinn Manual 2.0 on the International Law Applicable to Cyber Operations* (Cambridge, UK: Cambridge University Press, 2017), 1.

20. As Sanger has accurately described cyberweapons, "The weapons remain invisible, the attacks deniable, the results uncertain." David Sanger, *The Perfect Weapon: War, Sabotage, and Fear in the Cyber Age* (New York: Crown, 2018), xiv.

21. This is not the first time that non-state actors have played a potentially important role in the verification and enforcement of international rules. As one author has noted, "International NGO Landmine Monitor, with members in 95 countries, plays a major role in collecting information on violations of the Ottawa Convention. Though Landmine Monitor is not officially mentioned in the treaty, its findings are presented at the annual conference of states party to the agreement and have been used to present official allegations of treaty violations." Mark E. Donaldson, "NGOs and Arms Control Processes," in Williams and Votti, 199.

22. "About the Cybersecurity Tech Accord," Tech Accord, accessed November 14, 2018, https://cybertechaccord.org/about/.

23. Brad Smith, "The Price of Cyber-Warfare," April 17, 2018, RSA Conference, Moscone Center San Francisco, video, 21:11, https://www.rsaconference.com/events/us18/agenda/sessions/11292-the-price-of-cyber-warfare.

24. "Charter of Trust," Siemens, https://new.siemens.com/global/en/company/topic-areas/digitalization/cybersecurity.html.

25. Emmanuel Macron, "Forum de Paris sur la Paix: Rendez-vous le 11 Novembre 2018 | Emmanuel Macron," YouTube video, 3:21, July 3, 2018, https://www.youtube.com/watch?v=-tc4N8h hdpA&feature=youtube.

26. "Cybersecurity: Paris Call of 12 November 2018 for Trust and Security in Cyberspace," France Diplomatie press release, November 12, 2018, https://www.diplomatie.gouv.fr/en/french -foreign-policy/digital-diplomacy/france-and-cyber-security/article/cybersecurity-paris -call-of-12-november-2018-for-trust-and-security-in.

27. Ibid.

28. Charlotte Graham-McLay and Adam Satariano, "New Zealand Seeks Global Support for Tougher Measures on Online Violence," *New York Times*, May 12, 2019, https://www.nytimes .com/2019/05/12/technology/ardern-macron-social-media-extremism.html?searchResultPosi tion=1; Jacinda Ardern, "Jacinda Ardern: How to Stop the Next Christchurch Massacre," *New York Times*, May 11, 2019, https://www.nytimes.com/2019/05/11/opinion/sunday/jacinda-ardern -social-media.html?searchResultPosition=4.

29. Jeffrey W. Knopf, "NGOs, Social Movements, and Arms Control," in *Arms Control: History, Theory, and Policy, Volume 1: Foundations of Arms Control*, ed. Robert E. Williams Jr. and Paul R. Votti (Santa Barbara: Praeger, 2012), 174–75.

30. Ibid., 180.

31. Ibid.

32. The point here is not that the *Tallinn Manual* has been anything less than important. To the contrary, it has been critical. But it doesn't exactly have a "brand name" that sends a broad and succinct message at a time when public diplomacy needs to advance in an era dominated by social media.

33. Casper Klynge's Twitter account: Casper Klynge (@DKTechAmb), https://twitter.com/DK TechAmb.

34. Boyd Chan, "Microsoft Kicks Off Digital Peace Now Initiative to #Stopcyberwarfare," *Neowin*, September 30, 2018, https://www.neowin.net/news/microsoft-kicks-off-digital-peace-now -initiative-to-stopcyberwarfare; Microsoft, Digital Peace Now, https://digitalpeace.microsoft .com/.

35. Albert Einstein, "The 1932 Disarmament Conference," *Nation*, August 23, 2001, https://www .thenation.com/article/1932-disarmament-conference-0/.

CHAPTER 8: CONSUMER PRIVACY

1. European Union Agency for Fundamental Rights, *Handbook on European Data Protection Law, 2018 Edition* (Luxembourg: Publications Office of the European Union, 2018), 29.

2. Ibid., 30.

3. We made the call for federal legislation at a speech on Capitol Hill before the Congressional Internet Caucus. We called for a federal law to include four elements: a uniform baseline consistent with privacy laws around the world that would apply both online and offline; increased transparency for the collection, use, and disclosure of personal information; personal control over the use and disclosure of personal information; and minimum-security requirements for the storage and transit of personal information. Jeremy Reimer, "Microsoft Advocates the Need for Comprehensive Federal Data Privacy Legislation," *Ars Technica*, November 3, 2005, https://arstechnica.com/uncategorized/2005/11/5523-2/. For the original materials, see Microsoft Corporation, *Microsoft Advocates Comprehensive Federal Privacy Legislation*, November 3, 2005, https://news.microsoft.com/2005/11/03/microsoft-advocates-comprehensive-federal -privacy-legislation/; Microsoft PressPass, *Microsoft Addresses Need for Comprehensive Federal Data Privacy Legislation*, November 3, 2005, https://news.microsoft.com/2005/11/03/microsoft -addresses-need-for-comprehensive-federal-data-privacy-legislation/; video of Brad Smith at Congressional Internet Caucus, November 3, 2005, https://www.youtube.com/watch?v=Sj10rK DpNHE.

4. Martin A. Weiss and Kristin Archick, *U.S.-EU Data Privacy: From Safe Harbor to Privacy Shield* (Washington, DC: Congressional Research Service, 2016), https://fas.org/sgp/crs/misc/R44257 .pdf.

5. Joseph D. McClendon and Fox Rothschild, "The EU-U.S. Privacy Shield Agreement Is Unveiled, but Its Effects and Future Remain Uncertain," *Safe Harbor* (blog), Fox Rothschild, March 2, 2016, https://dataprivacy.foxrothschild.com/tags/safe-harbor/.

6. David M. Andrews, et. al., *The Future of Transatlantic Economic Relations* (Florence, Italy: European University Institute, 2005), 29; https://www.law.uci.edu/faculty/full-time/shaffer/pdfs /2005%20The%20Future%20of%20Transatlantic%20Economic%20Relations.pdf.

7. Daniel Hamilton and Joseph P. Quinlan, *The Transatlantic Economy 2016* (Washington, DC: Center for Transatlantic Relations, 2016), v.
8. An interesting contemporaneous account with Schrems as his case unfolded, see Robert Levine, "Behind the European Privacy Ruling That's Confounding Silicon Valley," *New York Times*, 9 Oct. 2015. https://www.nytimes.com/2015/10/11/business/international/behind-the-european-privacy-ruling-thats-confounding-silicon-valley.html.
9. Kashmir Hill, "Max Schrems: The Austrian Thorn in Facebook's Side," *Forbes*, February 7, 2012, https://www.forbes.com/sites/kashmirhill/2012/02/07/the-austrian-thorn-in-facebooks-side/#2d84e427b0b7.
10. Court of Justice of the European Union, "The Court of Justice Declares That the Commission's US Safe Harbour Decision Is Invalid," Press Release No. 117/15, October 6, 2015, https://curia.europa.eu/jcms/upload/docs/application/pdf/2015-10/cp150117en.pdf.
11. Mark Scott, "Data Transfer Pact Between U.S. and Europe Is Ruled Invalid," *New York Times*, October 6, 2015, https://www.nytimes.com/2015/10/07/technology/european-union-us-data-collection.html.
12. John Frank, "Microsoft's Commitments, Including DPA Cooperation, Under the EU-US Privacy Shield," *EU Policy Blog*, Microsoft, April 11, 2016, https://blogs.microsoft.com/eupolicy/2016/04/11/microsofts-commitments-including-dpa-cooperation-under-the-eu-u-s-privacy-shield/.
13. Grace Halden, *Three Mile Island: The Meltdown Crisis and Nuclear Power in American Popular Culture* (New York: Routledge, 2017), 65.
14. Julia Carrie Wong, "Mark Zuckerberg Apologises for Facebook's 'Mistakes' over Cambridge Analytica," *Guardian*, March 22, 2018, https://www.theguardian.com/technology/2018/mar/21/mark-zuckerberg-response-facebook-cambridge-analytica.
15. See Shoshana Zuboff, *The Age of Surveillance Capitalism: The Fight for a Human Future at the New Frontier of Power* (New York: PublicAffairs, 2019).
16. Julie Brill, "Millions Use Microsoft's GDPR Privacy Tools to Control Their Data — Including 2 Million Americans," *Microsoft on the Issues* (blog), Microsoft, September 17, 2018, https://blogs.microsoft.com/on-the-issues/2018/09/17/millions-use-microsofts-gdpr-privacy-tools-to-control-their-data-including-2-million-americans/.

CHAPTER 9: RURAL BROADBAND

1. "Wildfire Burning in Ferry County at 2500 Acres," *KHQ-Q6*, August 2, 2016, https://www.khq.com/news/wildfire-burning-in-ferry-county-at-acres/article_95f6e4a2-0aa1-5c6a-8230-9dca430aea2f.html.
2. Federal Communications Commission, *2018 Broadband Deployment Report*, February 2, 2018, https://www.fcc.gov/reports-research/reports/broadband-progress-reports/2018-broadband-deployment-report.
3. Jennifer Levitz and Valerie Bauerlein, "Rural America Is Stranded in the Dial-Up Age," *Wall Street Journal*, June 15, 2017, https://www.wsj.com/articles/rural-america-is-stranded-in-the-dial-up-age-1497535841.
4. Julianne Twining, "A Shared History of Web Browsers and Broadband Speed," NCTA, April 10, 2013, https://www.ncta.com/platform/broadband-internet/a-shared-history-of-web-browsers-and-broadband-speed-slideshow/.
5. Microsoft Corporation, *An Update on Connecting Rural America: The 2018 Microsoft Airband Initiative*, https://blogs.microsoft.com/uploads/prod/sites/5/2018/12/MSFT-Airband_Interactive PDF_Final_12.3.18.pdf.
6. Another problem with the FCC's approach is that it's "based on census blocks, which are the smallest geographic unit used by the US Census Bureau (although some are quite large—the biggest, in Alaska, is more than 8,500 square miles). If an internet service provider (ISP) sells broadband to a single customer in a census block, the FCC counts the entire block as having service." Ibid.
7. "Internet/Broadband Fact Sheet," Pew Research Center, February 5, 2018, https://www.pewinternet.org/fact-sheet/internet-broadband/.
8. Industry Analysis and Technology Division, Wireline Competition Bureau, *Internet Access Services: Status as of June 30, 2017* (Washington, DC: Federal Communications Commission, 2018), https://docs.fcc.gov/public/attachments/DOC-355166A1.pdf.

9. In 2018, we created a dedicated data science team to help us advance our work on key societal issues. We recruited one of Microsoft's most experienced data scientists, John Kahan, to lead the team. He had led a large team that applied data analytics to track and analyze the company's sales and product usage, and I had seen first-hand in weekly Senior Leadership Team meetings how this had improved our business performance. He also had a much broader set of interests, based in part on the work he and his team had pursued to use data science to better diagnose the causes of Sudden Infant Death Syndrome, or SIDS, to which John and his wife had lost their infant son, Aaron, more than a decade before. Dina Bass, "Bereaved Father, Microsoft Data Scientists Crunch Numbers to Combat Infant Deaths," *Seattle Times*, June 11, 2017, https://www.seattletimes.com/business/bereaved-father-microsoft-data-scientists-crunch-numbers-to-combat-infant-deaths/.

One of the first projects we gave to the new team was to dig into the concerns we had developed regarding the FCC's national data map on broadband availability. Within a few months, the team had used multiple data sets to analyze the broadband gap across the country, including data from the FCC and the Pew Research Center, as well as anonymized Microsoft data collected as part of ongoing work to improve the performance and security of our software and services. We published our initial conclusions in December 2018. Microsoft, "An Update on Connecting Rural America: The 2018 Microsoft Airband Initiative," 9. John and his team shared its findings with staff at the FCC and across the executive branch and provided demonstrations highlighting the data discrepancies in individual states by using a large Microsoft Surface Hub on Capitol Hill.

The team continued its work in 2019, including by asking the FCC and members of Congress to focus more on the issue. In April, we published specific recommendations that we believed would improve the accuracy of the FCC's data. John Kahan, "It's Time for a New Approach for Mapping Broadband Data to Better Serve Americans," *Microsoft on the Issues* (blog), Microsoft, April 8, 2019, https://blogs.microsoft.com/on-the-issues/2019/04/08/its-time-for-a-new-approach-for-mapping-broadband-data-to-better-serve-americans/. The same month, the Senate Committee on Commerce, Science, and Transportation zeroed in on the problem in a hearing. Committee Chairman Roger Wicker pointed out the deficiency in current data and said that "to close the digital divide we need to have accurate broadband maps that tell us where broadband is available and where it is not available at certain speeds." Mitchell Schmidt, "FCC Broadband Maps Challenged as Overstating Access," *The Gazette*, April 14, 2019, https://www.thegazette.com/subject/news/government/fcc-broadband-maps-challenged-as-overstating-access-rural-iowans-20190414. Jonathan Spalter, president and chief executive officer of the United States Telecom Association, said in the hearing that "the current yardstick, collecting data by census block, is inadequate. It means if a provider is able to serve a single location within that block then every location is considered served." Ibid.

10. Schmidt, "FCC Broadband Map."

11. "November 8, 2016 General Election Results," Washington Office of the Secretary of State, November 30, 2016, https://results.vote.wa.gov/results/20161108/President-Vice-President_By County.html.

12. "About the Center for Rural Affairs," Center for Rural Affairs, last updated 2019, https://www.cfra.org/about.

13. Johnathan Hladik, *Map to Prosperity* (Lyons, NE: Center for Rural Affairs, 2018), https://www.cfra.org/sites/www.cfra.org/files/publications/Map%20to%20Prosperity.pdf, 2, citing Arthur D. Little, "Socioeconomic Effects of Broadband Speed," Ericsson ConsumerLab and Chalmers University of Technology, September 2013, http://nova.ilsole24ore.com/wordpress/wp-content/ uploads/2014/02/Ericsson.pdf.

14. Ibid.

15. Jennifer Levitz and Valerie Bauerlein, "Rural America Is Stranded in the Dial-Up Age."

16. Ibid.

17. The FCC's universal service mechanism provides approximately $4 billion to landline carriers through Connect America Fund and legacy programs. In contrast, there is roughly $500 million available for wireless carriers through the Mobility Fund and legacy programs.

18. Sean Buckley, "Lawmakers Introduce New Bill to Accelerate Rural Broadband Deployments on Highway Rights of Way," Fiercetelecom, March 13, 2017, http://www.fiercetelecom.com/telecom/lawmakers-introduce-new-bill-to-accelerate-rural-broadband-deployments-highway-rights-way.

19. Microsoft Corporation, "United States Broadband Availability and Usage Analysis: Power BI Map," *Stories* (blog), Microsoft, December 2018, https://news.microsoft.com/rural-broadband/.
20. "Voice Voyages by the National Geographic Society," *The National Geographic Magazine*, vol. 29, March 1916, 312.
21. Ibid., 314.
22. Connie Holland, "Now You're Cooking with Electricity!" *O Say Can You See?* (blog), Smithsonian National Museum of American History, August 24, 2017, http://americanhistory.si.edu/blog/cooking-electricity.
23. Ibid.
24. "Rural Electrification Administration," Roosevelt Institute, February 25, 2011, http://rooseveltinstitute.org/rural-electrification-administration/.
25. Chris Dobbs, "Rural Electrification Act," *New Georgia Encyclopedia*, August 22, 2018, http://www.georgiaencyclopedia.org/articles/business-economy/rural-electrification-act.
26. "REA Energy Cooperative Beginnings," REA Energy Cooperative, accessed January 25, 2019, http://www.reaenergy.com/rea-energy-cooperative-beginnings.
27. "Rural Electrification Administration," Roosevelt Institute.
28. Ibid.
29. Rural Cooperatives, "Bringing Light to Rural America," March–April 1998, vol. 65, issue 2, 33.
30. "Rural Electrification Administration," Roosevelt Institute.
31. "REA Energy Cooperative Beginnings." REA Energy Cooperative.
32. Ibid.
33. Gina M. Troppa, "The REA Lady: A Shining Example, How One Woman Taught Rural Americans How to Use Electricity," *Illinois Currents*, https://www.lib.niu.edu/2002/ic020506.html.

CHAPTER 10: THE TALENT GAP

1. Jon Gertner, *The Idea Factory: Bell Labs and the Great Age of American Innovation* (New York: Penguin Press, 2012).
2. Brad Smith and Carol Ann Browne, "High-Skilled Immigration Has Long Been Controversial, but Its Benefits Are Clear," *Today in Technology* (blog), LinkedIn, December 7, 2017, https://www.linkedin.com/pulse/dec-7-forces-divide-us-bring-together-brad-smith/.
3. Brad Smith and Carol Ann Browne, "The Beep Heard Around the World," *Today in Technology* (blog), LinkedIn, October 4, 2017, https://www.linkedin.com/pulse/today-technology-beep-heard-around-world-brad-smith/.
4. Zapolsky acted quickly to mobilize Amazon's resources to support what became Washington State Attorney General Bob Ferguson's successful legal challenge to the first travel ban. Stephanie Miot, "Amazon, Expedia Back Suit Over Trump Immigration Ban," PCMag.com, January 31, 2017, https://www.pcmag.com/news/351453/amazon-expedia-back-suit-over-trump-immigration-ban. Monica Nickelsburg, "Washington AG Explains How Amazon, Expedia, and Microsoft Influenced Crucial Victory Over Trump," *Geekwire*, February 3, 2017, https://www.geekwire.com/2017/washington-ag-explains-amazon-expedia-microsoft-influenced-crucial-victory-trump/.
5. Jeff John Roberts, "Microsoft: Feds Must 'Go Through Us' to Deport Dreamers," *Fortune*, September 5, 2017, http://fortune.com/2017/09/05/daca-microsoft/.
6. Office of Communications, "Princeton, a Student and Microsoft File Federal Lawsuit to Preserve DACA," Princeton University, November 3, 2017, https://www.princeton.edu/news/2017/11/03/princeton-student-and-microsoft-file-federal-lawsuit-preserve-daca.
7. Microsoft Corporation, *A National Talent Strategy*, December 2012, https://news.microsoft.com/download/presskits/citizenship/MSNTS.pdf.
8. Jeff Meisner, "Microsoft Applauds New Bipartisan Immigration and Education Bill," *Microsoft on the Issues* (blog), Microsoft, January 29, 2013, https://blogs.microsoft.com/on-the-issues/2013/01/29/microsoft-applauds-new-bipartisan-immigration-and-education-bill/.
9. Mark Muro, Sifan Liu, Jacob Whiton, and Siddharth Kulkarni, *Digitalization and the American Workforce* (Washington, DC: Brookings Metropolitan Policy Program, 2017), https://www.brookings.edu/wp-content/uploads/2017/11/mpp_2017nov15_digitalization_full_report.pdf.
10. Ibid.

11. Nat Levy, "Q&A: Geek of the Year Ed Lazowska Talks UW's Future in Computer Science and Impact on the Seattle Tech Scene," *Geekwire,* May 5, 2017, https://www.geekwire.com /2017/qa-2017-geek-of-the-year-ed-lazowska-talks-uws-future-in-computer-science-and -impact-on-the-seattle-tech-scene/. Lazowska has been a tireless and effective champion for expanding access to computer science, including in higher education. He arrived at the University of Washington when it had only twelve computer science professors and Microsoft was a small start-up. As Bill Gates and Steve Ballmer led Microsoft to become a global technology leader, Lazowska played a decisive role in leading the University of Washington's work to establish one of the world's leading computer science programs. Both institutions have benefited from each other's success and a strong partnership between them, demonstrating in a dramatic way the symbiotic relationship that often exists between the tech sector and leading universities. See Taylor Soper, "Univ. of Washington Opens New Computer Science Building, Doubling Capacity to Train Future Tech Workers," *Geekwire,* February 28, 2019, https://www .geekwire.com/2019/photos-univ-washington-opens-new-computer-science-building -doubling-capacity-train-future-tech-workers/.

12. "AP Program Participation and Performance Data 2018," College Board, https://research.col legeboard.org/programs/ap/data/participation/ap-2018.

13. Ibid.

14. David Gelles, "Hadi Partovi Was Raised in a Revolution. Today He Teaches Kids to Code," *New York Times,* January 17, 2019, https://www.nytimes.com/2019/01/17/business/hadi-partovi -code-org-corner-office.html.

15. "Blurbs and Useful Stats," Hour of Code, accessed January 25, 2019, https://hourofcode.com /us/promote/stats.

16. Megan Smith, "Computer Science for All," https://obamawhitehouse.archives.gov/blog/2016 /01/30/computer-science-all.

17. "The Economic Graph," LinkedIn, accessed February 27, 2019, https://economicgraph.linke din.com/.

18. The Markle Foundation's Skillful initiative has led innovative work to develop skills-oriented hiring, training, and education efforts, based in part on work with LinkedIn. Steve Lohr, "A New Kind of Tech Job Emphasizes Skills, Not a College Degree," *New York Times,* June 29, 2017, https://www.nytimes.com/2017/06/28/technology/tech-jobs-skills-college-degree.html. After testing and validating successful efforts in Colorado, Skillful has expanded its work into Indiana. In a similar vein, Microsoft's subsidiary in Australia has worked with LinkedIn's Australian team and local governments to use LinkedIn data to better identify the skills that will most be in demand as the economy adopts more digital technology. Microsoft Australia, *Building Australia's Future-Ready Workforce,* February 2018, https://msenterprise.global.ssl.fastly .net/wordpress/2018/02/Building-Australias-Future-Ready-Workforce.pdf. The World Bank naturally is taking a global approach, working with LinkedIn to construct and validate metrics on skills, industry employment, and talent migration in over one hundred countries. Tingting Juni Zhu, Alan Fritzler, and Jan Orlowski, *Data Insights: Jobs, Skills and Migration Trends Methodology & Validation Results,* November 2018, http://documents.worldbank.org/curated/en /827991542143093021/World-Bank-Group-LinkedIn-Data-Insights-Jobs-Skills-and -Migration-Trends-Methodology-and-Validation-Results.

19. Paul Petrone, "The Skills New Grads Are Learning the Most," *The Learning Blog* (LinkedIn), May 9, 2019, https://learning.linkedin.com/blog/top-skills/the-skills-new-grads-are-learning -the-most.

20. I've served as the chair of the board of the Washington State Opportunity Scholarship program since it was created, having been appointed to the position first by Governor Christine Gregoire and then reappointed by Governor Jay Inslee.

21. Katherine Long, "Washington's Most Generous Scholarship for STEM Students Has Helped Thousands. Could You Be Next?" *Seattle Times,* December 28, 2018, https://www.seattletimes .com/education-lab/the-states-most-generous-scholarship-for-stem-students-has-helped -thousands-could-you-be-next/; Washington State Opportunity Scholarship, *2018 Legislative Report,* December 2018, https://www.waopportunityscholarship.org/wp-content/uploads/2018 /11/WSOS-2018-Legislative-Report.pdf.

22. Alan Greenspan and Adrian Wooldridge, *Capitalism in America: A History* (New York: Penguin Press, 2018), 393, citing Raj Chetty et al., "The Fading American Dream: Trends in Absolute

Income Mobility Since 1940," NBER Working Paper No. 22910, National Bureau of Economic Research, March 2017.

23. Brad Smith, Ana Mari Cauce, and Wayne Martin, "Here's How Microsoft and UW Leaders Want to Better Fund Higher Education," *Seattle Times*, March 20, 2019, https://www.seattle times.com/opinion/how-the-business-community-can-support-higher-education-funding/.

24. Ibid.

25. Hanna Scott, "Amazon, Microsoft on Opposite Ends of Tax Debate in Olympia," *MyNorthwest*, April 5, 2019, https://mynorthwest.com/1335071/microsoft-amazon-hb-2242-tax/.

26. Emily S. Rueb, "Washington State Moves Toward Free and Reduced College Tuition, With Businesses Footing the Bill," *New York Times*, May 8, 2019, https://www.nytimes.com/2019/05/08/education/free-college-tuition-washington-state.html.

27. Katherine Long, "110,000 Washington Students a Year Will Get Money for College, Many a Free Ride," *Seattle Times*, May 5, 2019, https://www.seattletimes.com/education-lab/110000-washington-students-a-year-will-get-money-for-college-many-a-free-ride/.

28. College Board, "AP Program Participation and Performance Data 2018," https://www.college board.org/membership/all-access/counseling-admissions-financial-aid-academic/number-girls-and-underrepresented.

29. "Back to School by Statistics," *NCES Fast Facts*, National Institute of Education Sciences, August 20, 2018, https://nces.ed.gov/fastfacts/display.asp?id=372.

30. Maria Alcon-Heraux, "Number of Girls and Underrepresented Students Taking AP Computer Courses Spikes Again," College Board, August 27, 2018, https://www.collegeboard.org/mem bership/all-access/counseling-admissions-financial-aid-academic/number-girls-and-underrepresented).

31. On the morning of August 5, 1888, Bertha Benz and her two teenage sons, Richard and Eugen, rolled the first patented horseless carriage, or Fahrzeug mit Gasmotorenbetrieb, onto the driveway ringing their home in Mannheim, Germany. Unbeknownst to her husband, Karl, Bertha was taking his three-wheeled contraption on a trip to her mother's home in Pforzheim—a sixty-mile journey that would later become known as the automobile's very first road trip. The trip wasn't easy. It took Bertha and her sons through steep and rugged terrain. They had to push the "smoking monster" up the muddy hills through Heidelberg to Wieslock and re-peatedly fill the engine with solvent purchased at local pharmacies. Dirty and exhausted, Bertha and her boys arrived at her mother's home that evening and telegraphed Karl to an-nounce their success. Their road trip made headlines, setting the stage for a new era of motor-ized transportation and the future success of the Mercedes-Benz motor company. Brad Smith and Carol Ann Browne, "The Woman Who Showed the World How to Drive," *Today in Technology* (blog), LinkedIn, August 5, 2017, https://www.linkedin.com/pulse/august-5-automobiles-first-road-trip-great-inventions-brad-smith/.

32. "Ensuring a Healthy Community: The Need for Affordable Housing, Chart 2," *Stories* (blog), Microsoft, https://3er1viui9wo30pkxh1v2nh4w-wpengine.netdna-ssl.com/wp-content/uploads/prod/sites/552/2019/01/Chart-2-Home-Price-vs.-MHI-1000x479.jpg.

33. Daniel Beekman, "Seattle City Council Releases Plan to Tax Businesses, Fund Homelessness Help," *Seattle Times*, April 20, 2018, https://www.seattletimes.com/seattle-news/politics/seat tle-city-council-releases-plan-to-tax-businesses-fund-homelessness-help/.

34. Matt Day and Daniel Beekman, "Amazon Issues Threat Over Seattle Head-Tax Plan, Halts Tower Construction Planning," *Seattle Times*, May 2, 2018, https://www.seattletimes.com/busi ness/amazon/amazon-pauses-plans-for-seattle-office-towers-while-city-council-considers-business-tax/.

35. Daniel Beekman, "About-Face: Seattle City Council Repeals Head Tax Amid Pressure From Businesses, Referendum Threat," *Seattle Times*, June 12, 2018, https://www.seattletimes.com/seattle-news/politics/about-face-seattle-city-council-repeals-head-tax-amid-pressure-from-big-businesses/.

36. "Ensuring a Healthy Community: The Need for Affordable Housing," *Stories* (blog), Micro-soft, https://news.microsoft.com/affordable-housing/.

37. "In 2015, about 57,000 people in the Seattle area endured commutes of at least ninety minutes from home to work, a jump of nearly 24,000 since 2010. That equals a 72 percent increase in just five years, ranking Seattle third among the 50 largest US metros for the rate of growth for mega-commuters." Gene Balk, "Seattle's Mega-Commuters: We Spend More Time Than Ever Traveling to Work," *Seattle Times*, June 16, 2017, https://www.seattletimes.com/seattle

-news/data/seattles-mega-commuters-we-are-spending-more-time-than-ever-traveling-to
-work/.

38. Brad Smith and Amy Hood, "Ensuring a Healthy Community: The Need for Affordable Housing," *Microsoft on the Issues* (blog), Microsoft, January 16, 2019, https://blogs.microsoft.com/on
-the-issues/2019/01/16/ensuring-a-healthy-community-the-need-for-affordable-housing/.

39. Paige Cornwell and Vernal Coleman, "Eastside Mayors View Microsoft's $500 Million Housing
Pledge with Enthusiasm, Caution," *Seattle Times*, January 23, 2019, https://www.seattletimes
.com/seattle-news/homeless/for-eastside-mayors-microsofts-500-million-pledge-for
-affordable-housing-is-tool-to-address-dire-need/.

40. Expanding low- and middle-income housing in the Seattle region will require a long-term effort, with no shortage of political and economic challenges. It took many years to dig the deep
housing hole and it will take many years for the region to dig its way out. As we recognized
within Microsoft when we made the decision to get involved, there undoubtedly will be days
when we'll need to work through some level of controversy, given the complexity of the issues.
But we felt it was important to get involved rather than stand on the sidelines and watch the
situation continue to deteriorate.

     One reason we were prepared to get involved was the leadership of former Washington
governor Christine Gregoire. After serving three terms as state attorney general and two as governor, she had the opportunity in 2013 to decide what to do next with her time and considerable
energy. We persuaded her to help us found and then serve as the CEO of Challenge Seattle to
bring together the region's largest companies to make a stronger civic contribution. Her commitment to the housing issue and her credibility across the region and political spectrum were instrumental in persuading us that it was a challenge we could help address in a meaningful way.
Information about Challenge Seattle can be found at https://www.challengeseattle.com/.

CHAPTER 11: AI AND ETHICS

1. Accenture, "Could AI Be Society's Secret Weapon for Growth? – WEF 2017 Panel Discussion," World Economic Forum, Davos, Switzerland, YouTube video, 32:03, March 15, 2017,
https://www.youtube.com/watch?v=6i_4y4lSC5M.

2. Asimov posited three laws of robotics. First, "A robot may not injure a human being or, through
inaction, allow a human being to come to harm." Second, "A robot must obey any orders given
to it by human beings, except where such orders would conflict with the First Law." And third,
"A robot must protect its own existence as long as such protection does not conflict with the
First or Second Law." Isaac Asimov, "Runaround," in *I, Robot* (New York: Doubleday, 1950).

3. In 1984–87, the focus was on advances in "expert systems" and their application to medicine,
engineering, and science. There were even special computers made and built for AI. This was
followed by a collapse and an "AI Winter," as it was called, for several years into the mid-1990s.

4. W. Xiong, J. Droppo, X. Huang, F. Seide, M. Seltzer, A. Stolcke, D. Yu, and G. Zweing, *Achieving Human Parity in Conversational Speech Recognition: Microsoft Research Technical Report
MSR-TR-2016-71*, February 2017, https://arxiv.org/pdf/1610.05256.pdf.

5. Terrence J. Sejnowski, *The Deep Learning Revolution* (Cambridge, MA: MIT Press, 2018), 31;
in 1986 Eric Horvitz coauthored one of the leading papers that made the case that expert systems would not be scalable. D.E. Heckerman and E.J. Horvitz, "The Myth of Modularity in
Rule-Based Systems for Reasoning with Uncertainty," *Conference on Uncertainty in Artificial
Intelligence*, Philadelphia, July 1986; https://dl.acm.org/citation.cfm?id=3023728.

6. Ibid.

7. Charu C. Aggarwal, *Neural Networks and Deep Learning: A Textbook* (Cham, Switzerland:
Springer, 2018), 1. The convergence of intellectual disciplines involved and affected by these
developments in recent decades is described in S.J. Gershman, E.J. Horvitz, and J.B. Tenen-
baum, *Science* 349, 273–78 (2015).

8. Aggarwal, *Neural Networks and Deep Learning*, 1.

9. Ibid., 17–30.

10. See Sejnowski for a thorough history of the developments that have led to advances in neural
networks over the past two decades.

11. Dom Galeon, "Microsoft's Speech Recognition Tech Is Officially as Accurate as Humans," Fu-
turism, October 20, 2016, https://futurism.com/microsofts-speech-recognition-tech-is-officially

-as-accurate-as-humans/; Xuedong Huang, "Microsoft Researchers Achieve New Conversational Speech Recognition Milestone," *Microsoft Research Blog*, Microsoft, August 20, 2017, https://www.microsoft.com/en-us/research/blog/microsoft-researchers-achieve-new-conversational-speech-recognition-milestone/.

12. The rise of superintelligence was first raised by I.J. Good, a British mathematician who worked as a cryptologist at Bletchley Park. He built on the initial work of his colleague, Alan Turing, and speculated about an "intelligence explosion" that would enable "ultra-intelligent machines" to design even more intelligent machines. I.J. Good, "Speculations Concerning the First Ultraintelligent Machine," *Advances in Computers* 6, 31–88 (January 1965). Among many other things, Good consulted for Stanley Kubrick's film *2001: A Space Odyssey*, which featured HAL, a famous runaway computer.

   Others in the computer science field, including at Microsoft Research, have been skeptical about the prospect of AI systems designing more intelligent versions of themselves or escaping human control based on their own thought processes. As Thomas Dietterich and Eric Horvitz suggest, "Such a process runs counter to our current understandings of the limitations that computational complexity places on algorithms for learning and reasoning." They note, "However, processes of self-design and optimization might still lead to significant jumps in competencies." T.G. Dietterich and E.J. Horvitz, "Rise of Concerns about AI: Reflections and Directions," *Communications of the ACM*, vol. 58, no. 10, 38–40 (October 2015), http://erichorvitz.com/CACM_Oct_2015-VP.pdf.

   Nick Bostrom, a professor at Oxford University, explored these issues more broadly in his recent book. Nick Bostrom, *Superintelligence: Paths, Dangers, Strategies* (Oxford: Oxford University Press, 2014).

   Within the computer science field, some use the term "singularity" differently, to describe computing power that grows so quickly that it's not possible to predict the future.

13. Julia Angwin, Jeff Larson, Surya Mattu, and Lauren Kirchner, "Machine Bias," *ProPublica*, May 23, 2016, https://www.propublica.org/article/machine-bias-risk-assessments-in-criminal-sentencing.

14. The article led to a lively debate about the definition of bias and how to assess the risk of it in AI algorithms. See Matthias Spielkamp, "Inspecting Algorithms for Bias," *MIT Technology Review*, June 12, 2017, https://www.technologyreview.com/s/607955/inspecting-algorithms-for-bias/.

15. Joy Buolamwini, "Gender Shades," Civic Media, MIT Media Lab, accessed November 15, 2018, https://www.media.mit.edu/projects/gender-shades/overview/.

16. Thomas G. Dietterich and Eric J. Horvitz, "Rise of Concerns About AI: Reflection and Directions," *Communications of the ACM* 58, no. 10 (2015), http://erichorvitz.com/CACM_Oct_2015-VP.pdf.

17. Satya Nadella, "The Partnership of the Future," *Slate*, June 28, 2016, http://www.slate.com/articles/technology/future_tense/2016/06/microsoft_ceo_satya_nadella_humans_and_a_i_can_work_together_to_solve_society.html.

18. Microsoft, *The Future Computed: Artificial Intelligence and Its Role in Society* (Redmond, WA: Microsoft Corporation, 2018), 53–76.

19. Paul Scharre, *Army of None: Autonomous Weapons and the Future of War* (New York: W. W. Norton, 2018).

20. Ibid., 163–69.

21. Drew Harrell, "Google to Drop Pentagon AI Contract After Employee Objections to the 'Business of War,'" *Washington Post*, June 1, 2018, https://www.washingtonpost.com/news/the-switch/wp/2018/06/01/google-to-drop-pentagon-ai-contract-after-employees-called-it-the-business-of-war/?utm_term=.86860b0f5a33.

22. Brad Smith, "Technology and the US Military," *Microsoft on the Issues* (blog), Microsoft, October 26, 2018, https://blogs.microsoft.com/on-the-issues/2018/10/26/technology-and-the-US-military/.

23. https://en.m.wikipedia.org/wiki/Just_war_theory; https://en.m.wikipedia.org/wiki/Mahabharata.

24. As we said, "To withdraw from this market is to reduce our opportunity to engage in the public debate about how new technologies can best be used in a responsible way. We are not going to withdraw from the future. In the most positive way possible, we are going to work to help shape it." Smith, "Technology and the US Military."

25. Ibid.

26. Adam Satariano, "Will There Be a Ban on Killer Robots?" *New York Times*, October 19, 2018, https://www.nytimes.com/2018/10/19/technology/artificial-intelligence-weapons.html.

27. SwissInfo, "Killer Robots: 'Do Something' or 'Do Nothing'?" *EurAsia Review*, March 31, 2019, http://www.eurasiareview.com/31032019-killer-robots-do-something-or-do-nothing/.

28. Mary Wareham, "Statement to the Convention on Conventional Weapons Group of Governmental Experts on Lethal Autonomous Weapons Systems, Geneva," Human Rights Watch, March 29, 2019, https://www.hrw.org/news/2019/03/27/statement-convention-conventional-weapons-group-governmental-experts-lethal.

29. Former US Marine Corps General John Allen, now president of the Brookings Institution, captured some of the critical ethical challenges eloquently when he wrote "From the earliest of times, humans have sought to restrain their baser instincts by seeking to govern them during the use of force: limiting its destructiveness and, in particular, the cruelty of its effects on innocents. These limits have been codified over time into a body of international law and professional military conduct that seeks to guide and limit the use of force and violence. And herein is the paradox: as we visit violence and destruction upon the enemy in war, we must do it with a moderation that acknowledges the necessity of its use, offering the means of discriminating between and among the participants, and admonishing us to apply proportionality." John Allen, foreword to *Military Ethics: What Everyone Needs to Know* (Oxford: Oxford University Press, 2016), xvi. See also Deane-Peter Baker, ed., *Key Concepts in Military Ethics* (Sydney: University of New South Wales, 2015).

30. Brad Smith and Harry Shum, foreword to *The Future Computed*, 8.

31. Oren Etzioni, "A Hippocratic Oath for Artificial Intelligence Practitioners," Tech Crunch, March 14, 2018, https://techcrunch.com/2018/03/14/a-hippocratic-oath-for-artificial-intelligence-practitioners/.

32. Cameron Addis, "Cold War, 1945–53," History Hub, accessed February 27, 2019, http://sites.austincc.edu/caddis/cold-war-1945-53/.

CHAPTER 12: AI AND FACIAL RECOGNITION

1. *Minority Report*, directed by Steven Spielberg (Universal City, CA: DreamWorks, 2002).

2. Microsoft Corporation, "NAB and Microsoft leverage AI technology to build card-less ATM concept," October 23, 2018, https://news.microsoft.com/en-au/2018/10/23/nab-and-microsoft-leverage-ai-technology-to-build-card-less-atm-concept/.

3. Jeannine Mjoseth, "Facial recognition software helps diagnose rare genetic disease," National Human Genome Research Institute, March 23, 2017, https://www.genome.gov/27568319/facial-recognition-software-helps-diagnose-rare-genetic-disease/.

4. Taotetek (@taotetek), "It looks like Microsoft is making quite a bit of money from their cozy relationship with ICE and DHS," Twitter, June 17, 2018, 9:20 a.m., https://twitter.com/taotetek/status/1008383982533259269.

5. Tom Keane, "Federal Agencies Continue to Advance Capabilities with Azure Government," *Microsoft Azure Government* (blog), Microsoft, January 24, 2018, https://blogs.msdn.microsoft.com/azuregov/2018/01/24/federal-agencies-continue-to-advance-capabilities-with-azure-government/.

6. Elizabeth Weise, "Amazon Should Stop Selling Facial Recognition Software to Police, ACLU and Other Rights Groups Say," *USA Today,* May 22, 2018, https://www.usatoday.com/story/tech/2018/05/22/aclu-wants-amazon-stop-selling-facial-recognition-police/633094002/.

7. While Amazon employees raised concerns in June 2018, the same month as Microsoft employees, Amazon did not respond directly to its workers until an internal meeting in November. Bryan Menegus, "Amazon Breaks Silence on Aiding Law Enforcement Following Employee Backlash," *Gizmodo,* November 8, 2018, https://gizmodo.com/amazon-breaks-silence-on-aiding-law-enforcement-followi-1830321057.

8. Drew Harwell, "Google to Drop Pentagon AI Contract After Employee Objections to the 'Business of War,'" *Washington Post*, June 1, 2018, https://www.washingtonpost.com/news/the-switch/wp/2018/06/01/google-to-drop-pentagon-ai-contract-after-employees-called-it-the-business-of-war/?noredirect=on&utm_term=.efa7f2973007.

9. Edelman, *2018 Edelman Trust Barometer Global Report*, https://www.edelman.com/sites/g/files/aatuss191/files/2018-10/2018_Edelman_Trust_Barometer_Global_Report_FEB.pdf.

10. Ibid., 30.

11. Kids in Need of Defense was founded in 2008 to provide children who are separated from their parents with pro bono legal counsel in immigration proceedings, https://supportkind.org/ten

-years/. Since its founding, KIND has trained more than 42,000 volunteers and now works with more than 600 law firms, corporations, law schools, and bar associations. It has become one of the largest pro bono legal organizations in the United States, and now pursues work in the United Kingdom as well. Wendy Young has led KIND since the first day it formally offered legal assistance to clients in 2009.

12. Annie Correal and Caitlin Dickerson, "'Divided,' Part 2: The Chaos of Reunification," August 24, 2018, in *The Daily*, produced by Lynsea Garrison and Rachel Quester, podcast, 31:03, https://www.nytimes.com/2018/08/24/podcasts/the-daily/divided-migrant-family-reunification.html.

13. Kate Kaye, "This Little-Known Facial-Recognition Accuracy Test Has Big Influence," International Association of Privacy Professionals, January 7, 2019, https://iapp.org/news/a/this-little-known-facial-recognition-accuracy-test-has-big-influence/.

14. Brad Smith, "Facial Recognition Technology: The Need for Public Regulation and Corporate Responsibility," *Microsoft on the Issues* (blog), Microsoft, July 13, 2018, https://blogs.microsoft.com/on-the-issues/2018/07/13/facial-recognition-technology-the-need-for-public-regulation-and-corporate-responsibility/.

15. Nitasha Tiku, "Microsoft Wants to Stop AI's 'Race to the Bottom,'" *Wired*, December 6, 2018, https://www.wired.com/story/microsoft-wants-stop-ai-facial-recognition-bottom/.

16. Eric Ries, *The Startup Way: How Modern Companies Use Entrepreneurial Management to Transform Culture and Drive Long-Term Growth* (New York: Currency, 2017), 96.

17. Brookings Institution, Facial recognition: Coming to a Street Corner Near You, December 6, 2018, https://www.brookings.edu/events/facial-recognition-coming-to-a-street-corner-near-you/.

18. Brad Smith, "Facial Recognition: It's Time for Action," *Microsoft on the Issues* (blog), December 6, 2018, https://blogs.microsoft.com/on-the-issues/2018/12/06/facial-recognition-its-time-for-action/.

19. We proposed that two steps be combined to make this effective. First, "Legislation should require tech companies that offer facial recognition services to provide documentation that explains the capabilities and limitations of the technology in terms that customers and consumers can understand." And second, "New laws should also require that providers of commercial facial recognition services enable third parties engaged in independent testing to conduct and publish reasonable tests of their facial recognition services for accuracy and unfair bias. A sensible approach is to require tech companies that make their facial recognition services accessible using the internet also make available an application programming interface or other technical capability suitable for this purpose." Smith, "Facial Recognition."

20. As we described it, new legislation should "require that entities that deploy facial recognition undertake meaningful human review of facial recognition results prior to making final decisions for what the law deems to be 'consequential use cases' that affect consumers. This includes where decisions may create a risk of bodily or emotional harm to a consumer, where there may be implications on human or fundamental rights, or where a consumer's personal freedom or privacy may be impinged." Smith, "Facial Recognition."

21. A camera that uses facial recognition at a specific location like an airport security checkpoint to help identify a terrorist suspect is one example. Even in this instance, however, it's important to require meaningful human review by trained personnel before a decision is made to detain someone.

22. *Carpenter v. United States*, No. 16-402, 585 U.S. (2017), https://www.supremecourt.gov/opinions/17pdf/16-402_h315.pdf.

23. Brad Smith, "Facial Recognition: It's Time for Action," *Microsoft on the Issues* (blog), December 6, 2018, https://blogs.microsoft.com/on-the-issues/2018/12/06/facial-recognition-its-time-for-action/.

24. As we pointed out, "The privacy movement in the United States was born from improvements in camera technology. In 1890, future Supreme Court Justice Louis Brandeis took the first step in advocating for privacy protection when he coauthored an article with colleague Samuel Warren in the *Harvard Law Review* advocating 'the right to be let alone.' The two argued that the development of 'instantaneous photographs' and their circulation by newspapers for commercial gain had created the need to protect people with a new right to privacy." Smith, "Facial Recognition," quoting Samuel Warren and Louis Brandeis, "The Right to Privacy," *Harvard Law Review*, IV:5 (1890), http://groups.csail.mit.edu/mac/classes/6.805/articles/privacy/Privacy_brand_warr2.html. As we pointed out, facial recognition is giving a new meaning to "instantaneous photographs" that Brandeis and Warren probably never imagined. Ibid.

25. Smith, "Facial Recognition."

26. One state legislator who took an interest in the idea was Reuven Carlyle, a Washington state senator who lives in Seattle and had worked in the tech sector before becoming a state legislator in 2009, https://en.wikipedia.org/wiki/Reuven_Carlyle. He wanted to champion a broad privacy bill, and he was interested in including facial recognition rules within it. Carlyle spent several months drafting his proposed legislation and talking with other state senators about its details. In part reflecting this effort, his bill, with new rules for facial recognition, gained the bipartisan support needed to pass out of the senate by a 46–1 vote in early March 2019. Joseph O'Sullivan, "Washington Senate Approves Consumer-Privacy Bill to Place Restrictions on Facial Recognition," *Seattle Times*, March 6, 2019, https://www.seattletimes.com/seattle-news/politics/senate-passes-bill-to-create-a-european-style-consumer-data-privacy-law-in-washington/.

27. Rich Sauer, "Six Principles to Guide Microsoft's Facial Recognition Work," *Microsoft on the Issues* (blog), December 17, 2018, https://blogs.microsoft.com/on-the-issues/2018/12/17/six-principles-to-guide-microsofts-facial-recognition-work/.

CHAPTER 13: AI AND THE WORKFORCE

1. "Last of Boro's Fire Horses Retire; 205 Engine Motorized," *Brooklyn Daily Eagle*, December 20, 1922, Newspapers.com, https://www.newspapers.com/image/60029538.

2. "1922: Waterboy, Danny Beg, and the Last Horse-Driven Engine of the New York Fire Department," *The Hatching Cat*, January 24, 2015, http://hatchingcatnyc.com/2015/01/24/last-horse-driven-engine-of-new-york-fire-department/.

3. "Goodbye, Old Fire Horse; Goodbye!" *Brooklyn Daily Eagle*, December 20, 1922.

4. Augustine E. Costello, *Our Firemen: A History of the New York Fire Departments, Volunteer and Paid, from 1609 to 1887* (New York: Knickerbocker Press, 1997), 94.

5. Ibid., 424.

6. "Heyday of the Horse," American Museum of Natural History, https://www.amnh.org/exhibitions/horse/how-we-shaped-horses-how-horses-shaped-us/work/heyday-of-the-horse.

7. "Microsoft TechSpark: A New Civic Program to Foster Economic Opportunity for all Americans," *Stories* (blog), accessed February 23, 2019, https://news.microsoft.com/techspark/.

8. Part of the inspiration for TechSpark was the political divide that emerged so dramatically in the 2016 US presidential election. The day after the election, in response to employee questions and requests, we did something we had not done before: We wrote a blog about our reaction to the presidential result. Brad Smith, "Moving Forward Together: Our Thoughts on the US Election," *Microsoft on the Issues* (blog), Microsoft, November 6, 2016, https://blogs.microsoft.com/on-the-issues/2016/11/09/moving-forward-together-thoughts-us-election/. One thing we noted was the way the political divide reflected an economic divide in the country, noting that "in a time of rapid change, we need to innovate to promote inclusive economic growth that helps everyone move forward." This led us to consider what more Microsoft could do to invest in efforts to promote technology-related economic growth outside the nation's largest urban centers, and on the two coasts.

    Under the leadership of Microsoft's Kate Behncken and Mike Egan, we founded the TechSpark initiative to pursue five strategies focused on six communities. We launched the program in Fargo in 2017 with North Dakota Governor and former Microsoft executive Doug Burgum. Brad Smith, "Microsoft TechSpark: A New Civic Program to Foster Economic Opportunity for all Americans," LinkedIn, October 5, 2017, https://www.linkedin.com/pulse/microsoft-techspark-new-civic-program-foster-economic-brad-smith/. TechSpark provides investments to expand computer science education in high schools, develop more pathways for people who wish to pursue new careers, broaden broadband availability, provide digital capabilities for the non-profit sector, and promote digital transformation across the local economy. https://news.microsoft.com/techspark/.

    The TechSpark team recruited and hired a local community engagement manager to lead the work in each of the six communities where it is investing. These are southern Virginia; northeastern Wisconsin; the area around El Paso, Texas, and across the border in Mexico; Fargo, North Dakota; Cheyenne, Wyoming; and central Washington. One of the strongest early investments involves a partnership with the Green Bay Packers across from Lambeau Field in Green Bay, Wisconsin. Microsoft and the Packers each committed $5 million to create TitleTownTech, which advances technology innovation in the region. Richard Ryman,

"Packers, Microsoft Bring Touch of Silicon Valley to Titletown District," *Green Bay Press Gazette*, October 20, 2017, https://www.greenbaypressgazette.com/story/news/2017/10/19/packers-microsoft-bring-touch-silicon-valley-titletown-district/763041001/; Opinion, "Titletown-Tech: Packers, Microsoft Partnership a 'Game Changer' for Greater Green Bay," *Green Bay Press Gazette*, October 21, 2017, https://www.greenbaypressgazette.com/story/opinion/editorials/2017/10/21/titletowntech-packers-microsoft-partnership-game-changer-greater-green-bay/786094001/.

9. Lauren Silverman, "Scanning the Future, Radiologists See Their Jobs at Risk," NPR, September 4, 2017, https://www.npr.org/sections/alltechconsidered/2017/09/04/547882005/scanning-the-future-radiologists-see-their-jobs-at-risk; "The First Annual Doximity Physician Compensation Report," *Doximity* (blog), April 2017, https://blog.doximity.com/articles/the-first-annual-doximity-physician-compensation-report.

10. Silverman, "Scanning the Future."

11. Asma Khalid, "From Post-it Notes to Algorithms: How Automation Is Changing Legal Work," NPR, November 7, 2017, https://www.npr.org/sections/alltechconsidered/2017/11/07/561631927/from-post-it-notes-to-algorithms-how-automation-is-changing-legal-work.

12. Radicati Group, "Email Statistics Report, 2015-2019," Executive Summary, March 2015, https://radicati.com/wp/wp-content/uploads/2015/02/Email-Statistics-Report-2015-2019-Executive-Summary.pdf.

13. Radicati Group, "Email Statistics Report, 2018–2022," March 2018, https://www.radicati.com/wp/wp-content/uploads/2017/12/Email-Statistics-Report-2018-2022-Executive-Summary.pdf.

14. Kenneth Burke, "How Many Texts Do People Send Every Day (2018)?" *How Many Texts People Send Per Day* (blog), Text Request, last modified November 2018, https://www.textrequest.com/blog/how-many-texts-people-send-per-day/.

15. Bill Gates, "Bill Gates New Rules," *Time*, April 19, 1999, http://content.time.com/time/world/article/0,8599,2053895,00.html.

16. Smith and Browne, "The Woman Who Showed the World How to Drive."

17. McKinsey Global Institute, *Jobs Lost, Jobs Gained: Workforce Transitions in a Time of Automation* (New York: McKinsey & Company, 2017), https://www.mckinsey.com/~/media/McKinsey/Featured%20Insights/Future%20of%20Organizations/What%20the%20future%20of%20work%20will%20mean%20for%20jobs%20skills%20and%20wages/MGI-Jobs-Lost-Jobs-Gained-Report-December-6-2017.ashx.

18. Ibid., 43.

19. Anne Norton Greene, *Horses at Work: Harnessing Power in Industrial America* (Cambridge, MA: Harvard University Press, 2008), 273.

20. "Pettet, Zellmer R. 1880–1962," WorldCat Identities, Online Computer Library Center, accessed November 16, 2018, http://worldcat.org/identities/lccn-no00042135/.

21. "Zellmer R. Pettet," *Arizona Republic*, August 22, 1962, Newspapers.com, https://www.newspapers.com/clip/10532517/pettet_zellmer_r_22_aug_1962/.

22. Robert J. Gordon, *The Rise and Fall of American Growth: The U.S. Standard of Living Since the Civil War* (Princeton, NJ: Princeton University Press, 2016), 60.

23. Ibid.

24. "Calorie Requirements for Horses," Dayville Hay & Grain, http://www.dayvillesupply.com/hay-and-horse-feed/calorie-needs.html.

25. Z.R. Pettet, "The Farm Horse," in U.S. Bureau of the Census, *Fifteenth Census, Census of Agriculture* (Washington, DC: Government Printing Office, 1933), 8.

26. Ibid., 71–77.

27. Ibid., 79.

28. Ibid., 80.

29. Linda Levine, *The Labor Market During the Great Depression and the Current Recession* (Washington, DC: Congressional Research Service, 2009), 6.

30. Ann Norton Greene, *Horses at Work: Harnessing Power in Industrial America* (Cambridge, MA: Harvard University Press, 2008).

31. Lendol Calder, *Financing the American Dream: A Cultural History of Consumer Credit* (Princeton, NJ: Princeton University Press, 1999), 184.

32. John Steele Gordon, *An Empire of Wealth: The Epic History of American Economic Power* (New York: HarperCollins, 2004), 299–300.

CHAPTER 14: THE UNITED STATES AND CHINA

1. Seattle Times Staff, "Live Updates from Xi Visit," *Seattle Times*, September 22, 2015, https://www.seattletimes.com/business/chinas-president-xi-arriving-this-morning/.

2. "Xi Jinping and the Chinese Dream," *The Economist*, May 4, 2013, https://www.economist.com/leaders/2013/05/04/xi-jinping-and-the-chinese-dream.

3. Reuters in Seattle, "China's President Xi Jinping Begins First US Visit in Seattle," *Guardian*, September 22, 2015, https://www.theguardian.com/world/2015/sep/22/china-president-xi-jinping-first-us-visit-seattle.

4. Julie Hirschfeld Davis, "Hacking of Government Computers Exposed 21.5 Million People," *New York Times*, July 9, 2019, https://www.nytimes.com/2015/07/10/us/office-of-personnel-management-hackers-got-data-of-millions.html.

5. Jane Perlez, "Xi Jinping's U.S. Visit," *New York Times*, September 22, 2015, https://www.nytimes.com/interactive/projects/cp/reporters-notebook/xi-jinping-visit/seattle-speech-china.

6. Evelyn Cheng, "Apple, Intel and These Other US Tech Companies Have the Most at Stake in China-US Trade Fight," *CNBC*, May 14, 2018, https://www.cnbc.com/2018/05/14/as-much-as-150-billion-annually-at-stake-us-tech-in-china-us-fight.html.

7. "Microsoft Research Lab—Asia," Microsoft, accessed January 25, 2019, https://www.microsoft.com/en-us/research/lab/microsoft-research-asia/.

8. Geoff Spencer, "Much More Than a Chatbot: China's XiaoIce Mixes AI with Emotions and Wins Over Millions of Fans," *Asia News Center* (blog), November 1, 2018, https://news.microsoft.com/apac/features/much-more-than-a-chatbot-chinas-xiaoice-mixes-ai-with-emotions-and-wins-over-millions-of-fans/.

9. "Microsoft XiaoIce, China's Newest Fashion Designer, Unveils Her First Collection for 2019," *Asia News Center* (blog), Microsoft, November 12, 2018, https://news.microsoft.com/apac/2018/11/12/microsofts-xiaoice-chinas-newest-fashion-designer-unveils-her-first-collection-for-2019/.

10. James Vincent, "Twitter Taught Microsoft's AI Chatbot to Be a Racist Asshole in Less Than a Day," *The Verge*, March 24, 2016, https://www.theverge.com/2016/3/24/11297050/tay-microsoft-chatbot-racist.

11. Richard E. Nisbet, *The Geography of Thought: How Asians and Westerners Think Differently . . . and Why* (New York: Free Press, 2003).

12. Henry Kissinger, *On China* (New York: Penguin Press, 2011), 13.

13. Ibid., 14–15.

14. Nisbett, *The Geography of Thought*, 2–3.

15. Ibid.

16. Microsoft has relied on ongoing conversations, partnerships, and memberships with key nongovernmental organizations to get an outside-in perspective on human rights issues. One group that has played a critical role in promoting a broader perspective and commitment to human rights across the tech sector is the Global Network Initiative, or GNI. Its membership combines human rights groups and tech companies that commit to a common set of principles and periodic auditing of our adherence to them. Global Network Initiative, "The GNI Principles," https://globalnetworkinitiative.org/gni-principles/. As noted by Guy Berger of the United Nations Education, Scientific, and Cultural Organization, or UNESCO, GNI is unique for its multi-stakeholder approach, given "its internal practice of bringing companies and civil society to dialogue." Guy Berger, "Over-Estimating Technological Solutions and Underestimating the Political Moment?" *The GNI Blog* (Medium), December 5, 2018, https://medium.com/global-network-initiative-collection/over-estimating-technological-solutions-and-underestimating-the-political-moment-467912fa2d20. As Berger recognized, GNI also plays an important "external role representing the common ground between these two constituencies to governments around the world." Ibid.

    Another place that has brought together the human rights and business communities is New York University's Center for Business and Human Rights at the Leonard N. Stern School of Business. Led by Michael Posner, one of the most respected human rights lawyers globally, the center focuses on the intersection of business and human rights, often by advancing pragmatic steps for companies to better address these challenges in their core operations. NYU Stern, "The NYU Stern Center for Business and Human Rights," https://www.stern.nyu.edu/experience-stern/about/departments-centers-initiatives/centers-of-research/business-and-human-rights.

17. He Huaihong, *Social Ethics in a Changing China: Moral Decay or Ethical Awakening?* (Washington, DC: Brookings Institution Press, 2015).

18. David E. Sanger, Julian E. Barnes, Raymond Zhong, and Marc Santora, "In 5G Race With China, U.S. Pushes Allies to Fight Huawei," *New York Times*, January 26, 2019, https://www.nytimes.com/2019/01/26/us/politics/huawei-china-us-5g-technology.html.
19. Sean Gallagher, "Photos of an NSA 'upgrade' factory shows Cisco router getting implant," ARS Technica, May 14, 2014, https://arstechnica.com/tech-policy/2014/05/photos-of-an-nsa-upgrade-factory-show-cisco-router-getting-implant/.
20. Reid Hoffman and Chris Yeh, *Blitzscaling: The Lightning-Fast Path to Building Massively Valuable Businesses* (New York: Currency, 2018).

CHAPTER 15: DEMOCRATIZING THE FUTURE

1. Kai-Fu Lee, *AI Superpowers: China, Silicon Valley, and the New World Order* (Boston: Houghton Mifflin Harcourt, 2018), 21.
2. Ibid., 169.
3. Ibid., 168–69.
4. "Automotive Electronics Cost as a Percentage of Total Car Cost Worldwide From 1950 to 2030," Statista, September 2013, https://www.statista.com/statistics/277931/automotive-electronics-cost-as-a-share-of-total-car-cost-worldwide/.
5. "Who Was Fred Hutchinson?," Fred Hutch, accessed January 25, 2019, https://www.fredhutch.org/en/about/history/fred.html.
6. "Mission & Facts," Fred Hutch, accessed January 25, 2019, https://www.fredhutch.org/en/about/mission.html.
7. Gary Gilliland, "Why We Are Counting on Data Science and Tech to Defeat Cancer," January 9, 2019, LinkedIn, https://www.linkedin.com/pulse/why-we-counting-data-science-tech-defeat-cancer-gilliland-md-phd/.
8. Ibid.
9. Gordon I. Atwater, Joseph P. Riva, and Priscilla G. McLeroy, "Petroleum: World Distribution of Oil," *Encyclopedia Britannica*, October 15, 2018, https://www.britannica.com/science/petroleum/World-distribution-of-oil.
10. "China Population 2019," World Population Review, accessed February 28, 2019, http://worldpopulationreview.com/countries/china-population/.
11. "2019 World Population by Country (Live)," World Population Review, accessed February 27, 2019, http://worldpopulationreview.com/.
12. International Monetary Fund, "Projected GDP Ranking (2018–2023)," Statistics Times, accessed February 27, 2019, http://www.statisticstimes.com/economy/projected-world-gdp-ranking.php.
13. Matthew Trunnell, unpublished memorandum.
14. Zev Brodsky, "Git Much? The Top 10 Companies Contributing to Open Source," WhiteSource, February 20, 2018, https://resources.whitesourcesoftware.com/blog-whitesource/git-much-the-top-10-companies-contributing-to-open-source.
15. United States Office of Management and Budget, "President's Management Agenda," White House, March 2018, https://www.whitehouse.gov/wp-content/uploads/2018/03/Presidents-Management-Agenda.pdf.
16. World Wide Web Foundation, *Open Data Barometer*, September 2018, https://opendatabarometer.org/doc/leadersEdition/ODB-leadersEdition-Report.pdf.
17. Trunnell, unpublished memo.
18. "Introduction to the CaDC," California Data Collaborative, accessed January 25, 2019, http://californiadatacollaborative.org/about.

CHAPTER 16: CONCLUSION

1. While a teenager at the Kentucky School for the Blind in Louisville, Anne Taylor decided that she wanted to learn computer science. It required that she spend half-days at the area's public high school, where she was the first student who was blind to study the subject. Anne discovered a passion that took her to Western Kentucky University, where she graduated with a degree in computer science. From there she went to work for the National Federation for the Blind, where she ultimately led the organization's team that advocates for accessibility across the tech industry. In 2015, Jenny Lay-Flurrie, Microsoft's chief accessibility officer, called

Anne and made an offer she couldn't refuse. "Come to Microsoft and work on the inside," Jenny encouraged. "See what impact you can have by working directly with our engineers to help shape product design before anything goes out the door."

2. Princeton University's Geniza Lab contains a massive trove of documents from Cairo's Ben Ezra Synagogue, including personal letters, shopping lists, and legal documents written in sacred Hebrew text that required a "dignified burial" in a special Ginza, or storage chamber. It's the largest known cache of Jewish manuscripts recorded. Scholars around the world have studied the artifacts since the late nineteenth century, and the job remains undone. By combining AI algorithms and computer vision to comb through thousands of digital fragments, Rustow's team has successfully matched fragments of the same document stored thousands of miles apart, matching tear shapes, bits of words, and the diameter of ink. When documents find their way "home" in this way, Rustow can finish painting a previously incomplete picture of how Jews and Muslims coexisted in 10th-century Islamic Middle East. AI has helped Rustow and her team of near eastern studies experts to accomplish in mere minutes what had been considered an insurmountable task. Robert Siegel, "Out of Cairo Trove, 'Genius Grant' Winner Mines Details of Ancient Life," NPR's *All Things Considered*, September 29, 2015, https://www.npr.org/2015/09/29/444527433/out-of-cairo-trove-genius-grant-winner-mines-details-of-ancient-life.

3. University of Southern California Center for Artificial Intelligence in Society, PAWS: Protection Assistant for Wildlife Security, accessed April 9, 2019, https://www.cais.usc.edu/projects/wildlife-security/.

4. Satya Nadella, "The Necessity of Tech Intensity in Today's Digital World," LinkedIn, January 18, 2019, https://www.linkedin.com/pulse/necessity-tech-intensity-todays-digital-world-satya-nadella/.

5. Einstein, "The 1932 Disarmament Conference."

6. Hoffman and Yeh, *Blitzscaling*.

7. This also requires the right type of leadership by a company's board of directors. Here too there is room for a broader approach at many tech companies. On the one hand, there's a risk that a board will defer so much to a strong and successful founder that it won't know enough about what's happening inside the company to ask the hard questions, or it will lack the courage even if these questions are apparent. On the other hand, a board that dives too deeply into some specifics may create confusion about the difference between the board's role in governing a company and a CEO's responsibility to lead and manage it.

At Microsoft, audit committee chair Chuck Noski has long focused on ensuring targeted but rigorous processes that go beyond financial controls and connect closely with the work of the internal audit team. Ironically, we also benefited when Judge Colleen Kollar-Kotelly decided in 2002 on her own initiative to approve our antitrust settlement subject to an additional obligation on the company's board of directors to create an antitrust compliance committee. A decade after that obligation expired, the board continues to rely on a regulatory and public policy committee led by former BMW CEO Helmut Panke to stay abreast of evolving issues for Microsoft. In addition to collaborating closely with the board's audit committee on issues like cybersecurity, this group once a year spends a day in an annual offsite where our management team reviews the past year's societal and political trends and we assess together our proactive work to address them. It's the type of exercise that forces us all to step back and look at the forest rather than just the trees, upping our game for the year ahead.

All this requires that directors have some real insight into a company's business, organization, people and issues. At Microsoft, our directors meet regularly in small groups with different sets of executives, participate in other meetings and attend the annual strategy retreat for our executive staff. At Netflix, where I'm a board member, CEO Reed Hastings arranges for directors to sit on a variety of staff meetings, large and small.

8. Margaret O'Mara, *The Code: Silicon Valley and the Remaking of America* (New York: Penguin Press, 2019), 6.

9. As O'Mara observes, the tech sector's "entrepreneurs were not lone cowboys, but very talented people whose success was made possible by the work of many other people, networks, and institutions. Those included the big-government programs that political leaders of both parties critiqued so forcefully, and that many tech leaders viewed with suspicion if not downright hostility. From the Bomb to the moon shot to the backbone of the Internet and beyond, public spending fueled an explosion of scientific and technical discovery, providing the foundation for generations of start-ups to come." Ibid., 5.

A similar phenomenon has long been noted by many public officials and lawyers in intellectual property fields. Despite resisting regulation, it's doubtful that tech companies would enjoy anything close to their hefty market valuations without the benefits of copyright, patent, and trademark laws, which have created the opportunity for inventors and developers to own the IP that they have created.

10. To the contrary, the reaction to the FAA's delegation of some regulatory certification to Boeing during the 737 MAX certification process has reflected official and public unease. The response quickly focused on requiring that the FAA base its assessment of the plane's safety fixes on additional outside review. Steve Miletich and Heidi Groover, "Reacting to Crash Finding, Congressional Leaders Support Outside Review of Boeing 737 MAX Fixes," *Seattle Times*, April 4, 2019, https://www.seattletimes.com/business/boeing-aerospace/reacting-to-crash-finding-congressional-leaders-support-outside-review-of-boeing-737-max-fixes/.

11. Ballard C. Campbell, *The Growth of American Government: Governance from the Cleveland Era to the Present* (Bloomington: Indiana University Press, 2015), 29.

12. Ari Hoogenboom and Olive Hoogenboom, *A History of the ICC: From Panacea to Palliative* (New York: W. W. Norton, 1976); Richard White, *Railroaded: The Transcontinentals and the Making of Modern America* (New York: W. W. Norton, 2011); Gabriel Kolko, *Railroads and Regulation: 1877–1916* (Princeton, NJ: Princeton University Press, 1965), 12.

13. Ibid.

14. "Democracy Index 2018: Me Too? Political Participation, Protest and Democracy," *The Economist* Intelligence Unit, https://www.eiu.com/public/topical_report.aspx?campaignid=Democracy2018.

Index

INDEX